About Island Press

Island Press is the only nonprofit organization in the United States whose principal purpose is the publication of books on environmental issues and natural resource management. We provide solutions-oriented information to professionals, public officials, business and community leaders, and concerned citizens who are shaping responses to environmental problems.

In 2000, Island Press celebrates its sixteenth anniversary as the leading provider of timely and practical books that take a multidisciplinary approach to critical environmental concerns. Our growing list of titles reflects our commitment to bringing the best of an expanding body of literature to the environmental community throughout North America and the world.

Support for Island Press is provided by The Jenifer Altman Foundation, The Bullitt Foundation, The Mary Flagler Cary Charitable Trust, The Nathan Cummings Foundation, The Geraldine R. Dodge Foundation, The Charles Engelhard Foundation, The Ford Foundation, The Vira I. Heinz Endowment, The W. Alton Jones Foundation, The John D. and Catherine T. MacArthur Foundation, The Andrew W. Mellon Foundation, The Charles Stewart Mott Foundation, The Curtis and Edith Munson Foundation, The National Fish and Wildlife Foundation, The National Science Foundation, The New-Land Foundation, The David and Lucile Packard Foundation, The Pew Charitable Trusts, The Surdna Foundation, The Winslow Foundation, and individual donors.

The
Local Politics
of
Global
Sustainability

The
Local Politics
of
Global
Sustainability

Thomas Prugh, Robert Costanza,
and Herman Daly

ISLAND PRESS
Washington, D.C. • Covelo, California

ISLAND PRESS is a trademark of The Center for Resource Economics.

Library of Congress Cataloging-in-Publication Data
Prugh, Thomas.
    The local politics of global sustainability / by Thomas Prugh, Robert Costanza, Herman E. Daly.
        p.   cm.
    Includes bibliographical references and index.
    ISBN 1–55963–743–9 (cloth) — 1–55963–744–7 (pbk.)
    1. Economic policy—Environmental aspects. 2. Environmental policy—Economic aspects. 3. Sustainable development—Political aspects. 4. Environmentalism—Political aspects. 5. Social ecology. 6. Political participation. 7. Local government.   I. Costanza, Robert.   II. Daly, Herman E.   III. Title.

HC79.E5 P747   2000
33.7'17—dc21   99–048263
                                                                              CIP

Only that day dawns to which we are awake.

—H. D. Thoreau

Contents

Preface

Hundreds of books have been written to bemoan our environmental dilemma and the damage we are doing to our planetary life-support system, and of the writing of them there seems no end. This book, however, will not add to that stream. It's not that we think those books are wrong; quite the contrary.

Broadly speaking, there are three camps offering analyses of our environmental problems; our position is closest to the third.

1. The Alfred E. Newman Camp. There are no problems, and there will be even fewer of them in the future.
2. The Technocratic Optimist Camp. Maybe there are a few little problems, but we can tinker here and there, install some compact fluorescent light bulbs, support mass transit, and get by. (Some optimists consider this position too gloomy. They think that economic growth, new technology, and human inventiveness in general will lead us inevitably on to the broad sunlit uplands of a forever better, richer, and more populous world.)
3. The Jeremiah Camp. We're in deep trouble, and getting out of it will require not merely new technology but also fundamental social, political, and economic transformation. Soon.

Most books about environmentalism have been written by members of Camp No. 3. We believe their analysis is the most likely to be true and also the least likely to be accepted.

One reason is that Jeremiahs are annoying. The biblical Jeremiah was stoned to death 2,600 years ago, and his prophetic heirs are not much better liked today. It is difficult to see the need for "fundamental social, political, and economic transformation" as anything but bad news, at least for those of us—mainly in the developed world (the North)—who are fairly comfortable with the status quo.

Moreover, it is easy to reject this message because it is plausibly deniable, at least for the moment. Many people (again, mainly northerners) have lived through several decades of environmental rabble-rousing, lawsuits, and legislation, and they have seen some progress. In the United States, for example, the air and water are now cleaner in some ways, and there is more land set aside for parks and wilderness areas. Some toxic emissions have been reduced. The Cuyahoga River has not caught fire since 1969. If profound crises await us, they tend to lurk rather than loom. As many polls attest, most people seem to have decided that the environment has been taken care of.

We do not believe it is quite yet time to prop up our feet and pour the margaritas. One bit of evidence is an important study published in 1995 by the Center for Economic and Security Alternatives.[1] The study assessed twenty-one environmental indicators and merged them into an index of trends for nine leading industrial nations that, as a group, have been the most environmentally active and have spent the most money. In the best case (Denmark), the index had declined nearly 11 percent since 1970. In the worst (France), it was down 41 percent. In the United States, the index declined 22 percent.

So we agree with those who argue that the need for change is more urgent than ever. Nevertheless, we do not view ourselves as committed members of the Jeremiah camp. We prefer to pitch our tents a little distance apart, because we believe that Jeremiah's legacy has been misunderstood and his name used falsely. Jeremiah's reputation rests on his passionate condemnation of what he saw as the evils of his society. Yet his most representative sentiment was not anger but grief, and his primary legacy was hope. He grieved for the failures of his people to be their best and held out the hope that social and religious atonement and reconciliation would create something like an ideal society. Some scholars regard him and the other prophets as the first utopians. Certainly, he offered a vision of a better future and—unusual for utopians—believed he knew how it could be attained.

We, too, regret that our culture seems largely heedless of trends that bode ill for our collective future. But we also think it obvious that although alarm is justified, something more is needed than novel ways to sound it. One of the shortcomings of the environmental movement has been its tendency toward reactivity, its heavy reliance on appeals to fear and anger, and its occasional obstructionism. Though these tactics remain useful, tactics are not a strategy. It is impossible to keep beating the gloom-and-doom drum forever; people will simply tune it out and eventually scorn the drummer. One warning sign for environmentalists is that they have now begun to appear in comic strips lumped together with other traditional Grinches, like bureaucrats and the Internal Revenue Service. When you become an object of satire rather than respect, it's time to take a look at your program.

In this book, we seek not merely to exhort people to stop practices that are bad but to help point the way toward an overarching Good. To do that, something positive is needed, a vision of a way of life that is attractive, ecologically harmonious, and supportable in the long term—in short, a way of life that is sustainable. In discussing one such vision, this book hopes to relieve the "concern fatigue" of those who are still receptive to the message but have become wearied by incessant doomsaying.

Our vision has to do more with process than product. More than a decade after the Brundtland Commission issued its famous definition of *sustainability,* the concept remains elusive and even treacherous. It is a mule than can be hitched to many wagons, and sometimes the mule is abused. Unlike mules, however, sustainability is a fertile concept that has generated many intellectual offspring, some of which do not seem to belong in the family. This book will talk about the importance of vision for sustainability, and will discuss some ideas about the ecological and economic constraints that should guide any effort toward sustainability. But we don't believe it is appropriate to prescribe a particular vision of a sustainable society. There are simply too many possibilities, and because they evolve over time and must be continuously selected among, prescription is pointless.

One of our main arguments is that the most important dimensions of sustainability are cultural and political. Later, we will discuss the idea that society has entered a phase of "postnormal science," which implies a different relationship between experts and all other stakeholders—at least in the sciences (ecology and economics) that concern us most here. The problems of sustainability—overconsumption, overpopulation, fossil fuel use, species destruction, and the rest—are not mainly technical. If they were, we could probably solve them in a few years. Nor do they affect simple linear systems, like a clock with a broken mainspring. The systems involved are complex and interactive in ways that make them inherently unpredictable.

As a result, the politics of communities' and nations' efforts to address their sustainability problems is much more important than any technical expertise they can muster. There are experts aplenty, but we cannot simply consult them for the "best" solutions, because nobody can know what those solutions are in any complete or final sense. The solutions must be explored and tested through a process of continuous adaptive learning. Deciding which options to try means making political choices that affect everyone and require wide support and engagement. A generation after its coinage, the slogan Power to the People takes on a new meaning.

Because there can be no permanent solutions in a world that is ecologically and culturally dynamic, these choices will have to be made again and again as circumstances evolve. Therefore, moving toward sustainability will require a radically broadened base of participants and a political process that continuously keeps them engaged. The process must encourage the perpetual hearing, testing, working through, and modification of competing visions *at the community level.*

Many readers may be surprised to discover this is rare (though, fortunately, not unheard of). The direct democratic process and structure we imagine are not widely used, even in the liberal democracies and obviously not in autocratic societies. To plant direct democracy in the latter will be difficult. Even to instill it in the former may seem unlikely in these times of low and declining voter turnout. We think it is possible, though, and in chapter 6 we discuss a few contemporary examples of direct democracy or its features. The key seems to be structuring political systems so that people's decisions matter. The system currently in use in the United States and other democracies does this rather badly. In representative democracies, citizens give away their power to elected representatives and to bureaucrats, who then make the decisions. When their decisions matter, people are more inclined to get involved and stay involved, and so create a powerful social expectation of involvement. As political scientist Benjamin Barber has noted, the ancient Greek term for the politically uninvolved person was *idiot*. Our first choice is therefore to be either citizens or idiots.

Once this process of involvement is established, sustainability becomes a moving beacon drawing us onward, not a predefined goal whose achievement marks the end of the journey. Apart from a few general characteristics—equitability, ecological sensitivity, democratic politics (autocracies are probably not sustainable, for reasons we will discuss later)—we cannot describe the sustainable society in advance. We can, however, say something about what it is not. Following the above, a sustainable society is not utopia, if that means a static "place, state, or situation of ideal perfection," as the dictionary puts it. It is more useful to think in terms of *genotopia*: the place that is continually and purposefully unfolding, emerging, or being reborn.

Because we are not talking about the perfect society, there is no pressure to assume perfect humans. Strictly speaking, sustainability does not require human virtue at all. If we fail to achieve sustainability, nature will impose it; but we would probably prefer the version we choose. Just as it is fatuous to speak of "saving the planet" (the planet will be here long after we're fossils), we should be clear about what we are trying to sustain: a society that works for us and our descendants ecologically, economically, morally, culturally, and politically. Sustainability on these terms *will* require certain virtues, especially restraint and the ability to learn adaptively.

One final thing sustainability is not. Despite appearances, it is not primarily global. To be sure, the world's the stage; a sustainable community or nation surrounded by unsustainable neighbors is a brave failure. Moreover, most issues of sustainability transcend national borders and will require international cooperation to address.

But we believe communities are the primary locus of responsibility for creating a sustainable world. The admonition to Think Globally, Act Locally retains its wisdom despite years of bumper-sticker overexposure. Directed sustainability will come about in neighborhoods or not at all. Humans seem evolved for communities of manageable size, and most of the individual behaviors and attitudes that support sustainability are best nurtured at the community level. The political structure and process necessary for a regionally, nationally, and globally sustainable society must be built on a foundation of local communities. To that end, this book will explore some ideas about community politics conducive to sustainability.

The late Isaiah Berlin once said something to the effect that thought is like money: if you've made your own, it must be counterfeit. We freely admit to being only kleptomanic tourists in regions that have been explored by countless political thinkers for hundreds or thousands of years. The ideas we briefly present here come from many places and are neither novel nor original, but we believe they are newly relevant to the sustainability and character of human society.

Having made that general acknowledgment, we would also like to thank several people for important contributions to this book, beginning with Todd Baldwin, our editor at Island Press, who worked with us patiently over many months to help develop this book into a coherent statement of ideas (of course, we alone are responsible if it does not seem so to our readers). We also owe a special debt of thanks to people who offered information, insights, suggestions, and support during the writing of the book, among them Susan Anderson, Ingrid Dankmeyer, Angus Duncan, Arthur Dye, John Fregonese, William Frost, Susan Hanna, Peter May, Richard May Jr., Thomas McCarthy, Michael McGinnis, Steven Scott, Ethan Seltzer, Zack Semke, Rachel Shimshak, Mary Lou Soscia, Lisa Speckhardt, Theresa Trump, Bettina von Hagen, Karl Weist, Ted Wolf, Pat Wortman, and several anonymous reviewers.

Note

1. G. Alperovitz et al., *Index of Environmental Trends: An Assessment of Twenty-One Key Environmental Indicators in Nine Industrialized Countries over the Past Two Decades* (Washington, DC: National Center for Economic and Security Alternatives, 1995).

Chapter 1
Introduction

What is worth saving? What forms of wealth are truly valuable? What, if anything, should be preserved today for future generations to enjoy, long after those who must decide are gone? These are the questions of sustainability, and every living human being daily answers them in a thousand ways, consciously or not. Achieving sustainability primarily means bringing these questions, and how we answer them, into the forefront of consciousness.

For serious environmentalists, sustainability issues are already uppermost. Environmentalists have long sought to protect certain features of the world—wilderness, other species, clean water, and so on—that they believed were important in the long term, as well as in the here and now. But the circumstances underlying the issues are changing, even for them. For a long time, economics figured little in the choices of what to protect. The human economy was small enough that it could operate as though it were separate from the environment. Today, however, humans making their daily livings extend their influence into every part of the natural world. Humans can hardly make any economic decision, profound or trivial, that has no effect on the environment. China's mammoth Three Gorges hydroelectric dam project (estimated cost: up to $79 billion) will flood nearly 1,500 cities, towns, and villages and vast expanses of prime farmland while displacing at least 1.2 million people.[1] At the other end of the scale, leaving the bathroom lights on all night long can add a pound or two of carbon dioxide to the atmosphere that otherwise might have remained locked up in some electric utility's coal,[2] and perhaps the globe warms a little faster as a result. As someone once said, we can never do just one thing. For instance, rabies has recently been eliminated in Euro-

1

pean foxes, but at the cost of an explosion in the fox population and the rise of other vulpine diseases, including a tapeworm that can infect and kill people.[3]

Large or small, every act counts. This is a relatively new condition, and one we are not very well prepared to handle. The global human economy, driven by increases in population and wealth, has ballooned to the point where humans are crowding out, or controlling for their own use, a large and rapidly growing fraction of the natural world's renewable resource output.[4] As a result, decisions about economic activity and environmental preferences have become profoundly interdependent. To make a decision in one arena is to make one in the other as well. A century ago, a government could set aside a few hundred square miles of wilderness with few economic repercussions. There was always plenty more land elsewhere. Now, with so much of the natural world stressed by human demand, there is not always "plenty more," of land or any other resource. It's not even clear there is enough. Today, proposals to create wilderness areas always trigger rancorous debate.

The strains we are imposing on the natural world mean that more and more decisions about it are guided by economics rather than by other considerations. The luxury of deciding to preserve some feature of the natural environment for any noneconomic reason—because it is beautiful or sacred, because it holds scientific interest, or simply because its existence and the fact that humans did not make it gives it intrinsic worth—is fast disappearing. Increasingly, we must preserve the natural environment because we need it to support our lifestyles and to ensure that we and our descendants will have some choice about the kind of lives we will lead.

So now the talk is about "sustainability." Because that is what this book is about, it ought not go any further without attempting a definition. The trouble is, sustainability is a big, sloppy term for a big, complex subject. Obviously, a thing or activity is sustainable if it can be kept going for a long time. To extend the definition to the environmental sustainability of human culture is more complicated. The easiest way is to start with what sustainability is not.

First, it is not about the survival of humanity as a species. "People are inexterminable—like flies and bedbugs," Robert Frost said. Strictly speaking, he was probably wrong; in the very long term, extinction is most species' fate and it may be humanity's, too. Short of a cataclysmic collision with a meteor, however, no plausible combination of ecological setbacks would completely wipe out *Homo sapiens sapiens,* as we optimistically call ourselves. Even serious environmental degradation or climate change would probably "only" lead to a much-reduced carrying capacity. That would result in many millions of premature deaths and the economic and cultural impoverishment of the survivors, and would be a terrible fate (all the more so for being avoidable). But the human species would continue. Humans have evolved over eons marked by repeated radical environmental change and appear to be not only well adapted to change but fundamentally formed by it.[5]

Likewise, the existence of the biosphere is not in much danger, either, and that is not what "saving the planet" should be taken to mean. The biosphere has survived billions of years of evolution and upheaval. However, life's forms and expressions show a great deal of changeability. The fossil record is replete with evidence of epochal waves of biological transformation that have marched repeatedly across the face of the planet. Antarctica, for example, apparently only became the cruel icebox of popular perception about 15 million years ago; before that time, the ice came and went, alternating with vast forests.[6]

The more scientists study the interactions of living creatures and the inanimate world around them, the more complex, dynamic, and unexpected those interactions appear. The familiar idea of a "balance of nature" has mostly been superseded by that of perpetual change. An ecosystem can appear static in the short run but can shift among multiple states over longer periods, depending on climatic and geological forces and the distribution of the plants, animals (including humans), and microorganisms that constitute its membership. Humans may find that ecosystems are varyingly hospitable as a result of these shifts, because it is easier to make a living in some than in others. But ecosystems rarely collapse and disappear like bursting bubbles and the global biosphere is unlikely to do so either.

If sustainability is not about humans or life in general dying out, what is it about? Judging by some of the hundreds of published definitions and commentaries, anything you like. As the examples below suggest, the concepts of sustainability and sustainable development have been stretched to the snapping point by attempts to make them fit many agendas. Definitions cover the entire spectrum. Some are straightforwardly technical:

> [Sustainability is] the idea of organizing an economic system so that it produces an enduring flow of output. . . . It is both the output of the economy that needs to be sustained, *and* the underlying resource base that gives rise to that output.[7]

Others are blunt:

> To put it crassly, consumers want consumption sustained. Workers want jobs sustained. . . . Sustainability calls to and is being called for by many, from tribal peoples to the most erudite academics, from peasant farmers to agroindustrialists, from denim-clad eco-activists to pinstripe-suited bankers. With the term meaning something different to everyone, the quest for sustainable development is off to a cacophonous start.[8]

Some aim for loftiness:

> Sustainability is a process with a beginning but no end; and in considerable measure it is a social construct. It requires . . . recognizing and respecting ecological integrity. It also requires a human vision for nature's duration, on terms hospitable to us and millions of other creatures. It is thus a matter of human imagination and dreaming as well as concrete technologies, tasks and policies.[9]

When politicians enter the fray, the definitions can become self-serving:

> Sustainable development . . . is a way to fulfill the requirements of the present without compromising the future. When policies of sustainable development are followed, our economic and our environmental objectives are both achieved. In fact, America's entire approach to bilateral and multilateral assistance is based on the concept of sustainable development.[10]

Others are ready with cautions:

> The synthesis of environmental and economic imperatives
> popularly called "sustainable development" must become a
> reality, not just a slogan.[11]

The definition that essentially started the whole discussion of sustainable development, from the 1987 Brundtland Commission report, remains one of the simplest: "Sustainable development is . . . development that meets the needs of the present without compromising the ability of future generations to meet their own needs."[12] Of course, like the others, it is easy to say.

All concepts of sustainability share the element of longevity: everybody wants something to persist. Beyond that, they all conceal major ambiguities and difficulties. Even the most specific ones—"organizing an economic system so that it produces an enduring flow of output," for example—leave important questions unanswered. How is the organizing to be achieved? What qualifies as "enduring"—fifty years or five hundred? What kind of output—Mercedes and Lexus cars, or millet and lentils? Who gets it—the poor who need it but cannot afford it, or the rich and middle class who can pay? Who decides? Can an economy be sustainable if it is unfair, or is viewed as unfair? That is, must sustainability include a social justice dimension?

If sustainability is only about meeting our *needs,* those are minimal. What about our *wants?* Nearly 90 percent of economic activity in rich nations goes toward satisfying wants,[13] and those seem insatiable. "Wants" have a way of becoming "needs": try living without a telephone in a culture where everyone else has one. How many of our wants can we expect to satisfy? How can we know what future generations will need or want? We cannot know—so we must somehow figure out what our obligations are. Should we leave them a specific collection of natural and "built" resources, a lot of money, or just our best wishes?

These difficult questions are not essentially technical, but questions of values. They cannot be answered by simply asking the experts. Sustainability will be achieved, if it all, not by engineers, agronomists, economists, and biotechnicians but by citizens.

To be sure, sustainability truly has an important technical dimension (which will be explored in some detail in chapter 2). A sustainable world of any reasonable definition will need a minimum supply of natural capital (resources). For several reasons, that threshold is difficult to locate, and it would be prudent to act as though we were already approaching it. Therefore, humanity should be taking certain steps of a technical nature to preserve its social choices in the decades ahead. For instance, with fossil fuels threatening to run out or become dangerous to use in the near future, it makes sense to work harder at developing renewable substitutes and more efficient machines.[14]

But workable answers to questions like those posed above, which concern shared, public values, must be found in the political arena, and that is why the most important dimension of sustainability is political. Until the value questions are identified and addressed through a political process, answering the technical questions is only potentially useful. The technical questions have to do with how to move toward a goal, but the political questions have to do with what goal is to be sought and the roles of the various players.

Late in 1997, newspapers were full of stories about the summit conference on global warming held in Kyoto, Japan, in December. They repeated a familiar, three-part refrain about environmental problems:

- The science is uncertain—fairly clear on some important points but not on others.
- Despite that, many things, technological and otherwise, could be done to attack the problem.
- An adequate response, however, would cost somebody a lot of money, at least in the short term (though it might save money in the long run).

This is a potent blend of facts. Human-caused change in the global climate is probably now unavoidable, but it could be minimized. Earth's climate seems to be warming because human economic activity (mainly combustion of coal, oil, and natural gas) is adding to the gases in the atmosphere that help trap the sun's heat and keep the global average temperature at about 60 degrees Fahrenheit. Hundreds of technical possibilities exist for reducing and eliminating dependence on fossil fuels and converting human energy systems to run on renewable resources. Moreover, there is no need to wait for the emergence of hypothetical technologies down the road; given the political will, we could make major progress with existing or rapidly developing technology in only a few years.

For many people, the evidence of global warming and the availability of compact fluorescent light bulbs and high-mileage automobiles add up to a compelling argument for urgent action to reduce emissions of green-house gases. But the high short-term costs create an opposing con-stituency, which claims that taking action will cause too much economic pain by retarding growth and trade.[15] (This argument was prominent at the Kyoto conference, which coincidentally opened while Asia was reeling from an economic crisis involving currency devaluations, stock market collapses, and the shocking bankruptcies of several large corporations.) The uncertainty of the science, too, makes it easy to sow doubt. If Earth is warming, skeptics argue, it might be only the upswing phase of a normal climate fluctuation. If it is warming artificially, how can we know for cer-tain that fossil fuels and deforestation are the culprits, since the pattern of increases in concentrations of atmospheric carbon dioxide does not track with the apparent warming seen so far? Even if Earth continues to heat up, nobody can say for sure how much; maybe only a degree or two. A little warming wouldn't necessarily be a bad thing, because growing seasons would lengthen in many areas, and some crops would do better with higher concentrations of carbon dioxide in the air. Besides, the skeptics conclude, environmentalists are always warning of global eco-disaster.

And so on. Expand this scenario to include all the other environmen-tal concerns facing the world, and the stage is set for broad and multifac-eted conflict over sustainability.

Sustainability thus brings people and governments face-to-face with the rawest political condition: we must decide to do something, or decide not to decide, yet we differ about courses of action and even about the facts, and we do not have a universally accepted ultimate authority, like science or religion, from which to seek the answers. How to proceed?

There is only one response to this question, and that is politics. Because the conflict is about values, sustainability must be socially and politically defined. Sustainability is provisional; it is subject to multiple conceptions and continuous revision, the very stuff of politics.

Like most political issues, sustainability is also a Trojan horse: in this case, for issues of economic, environmental, and social justice and equity. What sustainability is really about is the scope, quality, richness, and benignity of human culture, the biosphere and the economic life we make from them, and the distribution of those benefits, both now and over time. "Saving the planet" really means preserving and fairly allocating the ability of present and future generations to fashion societies that make life worth living. That is difficult enough in good times and is likely to become even more difficult if the environment continues to deteriorate (see Robert Kaplan's dismal scenario, discussed in chapter 3, for one plausible future). Even the structure of our societies is not assured, including those of the resource- and culture-rich nations in the developed world. Just as most species have flourished for a while and then disappeared from Earth, *so have most complex societies,* including the Western Chou empire of ancient China, the Harappans of India, several Mesopotamian empires, Old Egypt, the Hittites, the Minoans, the Mycenaeans, the Western Roman empire, the Lowland Classic Mayan empire, the Chacoans of the southwestern United States, and many others.[16] The irony of Shelley's famous poem "Ozymandias"—"Look on my works, ye Mighty, and despair!"—still applies. As has often happened in the past, environmental ignorance and mismanagement could force a great simplification in human societies.

———

If sustainability must be addressed first and foremost through politics, how are we doing?

A great deal of political activity centers around sustainability issues. Nationally and internationally, negotiations are under way on saving whales and preserving declining ocean fisheries. The United Nations runs its Environment Program. The Global Environment Facility is available to help developing countries address environment concerns. In many developed nations, the environment is often the subject of contentious legislation to prevent species extinction or to allow easier resource extraction. In the United States, Vice-President Gore achieves fame and notoriety by writing an impassioned pro-environment book, *Earth in the Balance*—and then struggles to save his political career from the charge of environmental extremism. Politically, almost everywhere, the environment counts.

It does not count nearly as much as growth and trade, however. As the Kyoto conference illustrated, short-term economics colors everything. News reports said that the chances of a "breakthrough agreement" on climate change were seriously threatened by Asia's economic woes.

Moreover, little of the political activity centered on the environment takes place where it matters most. The questions of sustainability are debated gingerly in the highest councils of government, but in the arena of everyday life, where ordinary people make billions of daily decisions that shape the common future, hardly a word is heard on the subject. Neighbors don't hang over the backyard fence discussing the odds of a 5-degree warming and how it might change life for their grandchildren—not even in the rich, industrial nations of the West, which have the most leverage and the greatest responsibility for climate change. There is nothing resembling a true national discussion of that or any other sustainability issue. Elected leaders, endlessly trolling the seas of public opinion for signs of major currents of thought, can hardly be blamed for detecting little support for serious action on sustainability—hence, the tentativeness of such conferences as the Kyoto gathering. Our politics simply fails to engage most people in such questions.

However, our political systems could be reformed to make them more engaging and responsive. One key element is to arrange the systems so people's political decisions matter.

Anyone active in his or her local civic association understands this. It became vividly clear to one of the authors when he became embroiled in a battle over speed humps on his neighborhood's major through street. Because of mishandling, the association meeting at which the issue was to be decided was held with little notice and gave the appearance of an attempt to sneak through a pro–speed-hump vote. Backed against a wall, angry opponents went from house to house with flyers claiming (wrongly, in supporters' view) that speed humps would depress property values and fatally delay emergency vehicles.

Most meetings of this association typically attracted fifteen or twenty faithful members, in a neighborhood of several thousand. The night of the speed-hump vote, nearly two hundred people jammed the local recreation center, filling every seat, lining the walls, and spilling out the door.

They shouted at each other late into the night. The exchanges were chaotic and exciting, often angry and never elegant, but it was democracy in action. And it made a difference: opponents fought the issue to a draw. In the end, the county installed speed humps on only one block of a half-mile street.

Issues that hit close to home engage people. Even in democracies, though, the political system tends to sever people from the issues by handing over the power to resolve them to elected representatives and non-elected public officials. Representation is so normal in developed nations that it is taken for granted, even seen as the only way to run a very large democracy. But it is a Faustian bargain. It relieves ordinary people of the details of governance, allowing them to shuck the mantle of citizen and embrace the role of consumer. The price is that they give away their power to choose directly what their communities will be like.

We need a politics of engagement, not a politics of consignment. A more engaging politics will be necessary to achieve a sustainable world of our choice, as opposed to one imposed by nature's unpredictable responses to abuse. Chapter 4 will set the stage for this discussion by reviewing the excesses of liberal democratic capitalism and asking whether the model of human psychology upon which it rests is inevitable. The short answer is no, which opens the door to the possibility of politically refashioning society for sustainability. Chapter 5 then explores some of the political issues affecting sustainability, with a particular focus on one option for reform, a type of self-governance that political scientist Benjamin Barber calls "strong democracy."[17] Strong democracy offers several immediate advantages over current systems:

- It would make communities stronger and more reflective of their residents' visions for their common lives. Strong democracy builds community by engaging people with each other as they struggle to address common issues. It strengthens the "us" without sacrificing the "me."
- Strong democracy disperses power, redistributing it downward so that governance is less susceptible to dominance by special interests.
- Strong democracy acts as a reality check by bringing citizens more directly into contact with the problems of governance.

These advantages, valuable in themselves, also serve the cause of sustainability. Because strong democracy is "politics understood as the creation of a vision that can respond to and change with the changing world,"[18] it is precisely in tune with the requirements of sustainability. As later chapters will show, these include: (1) vision, (2) a broader stakeholder base and wider citizen engagement, (3) tolerance for pluralism, and (4) adaptability to changing circumstances and values. Strong democracy thus appears better equipped than current systems to set us on a sustainable path to the future. A change of direction seems urgent, or else, as the saying goes, we will end up where we are going. That change will have to be made politically because politics is "the means by which men are emancipated from determinative historical forces."[19] If we are to steer away from the impending future of too many billions of people living subsistence lives on a hotter, ecologically degraded planet, we must transform our politics.

The book concludes by shifting focus from the general to the particular. In chapter 6, we will look at historical instances of vigorous, direct democratic cultures and sample some modern efforts to revive strong democratic governance, then in chapter 7 try to gauge the prospects for such a revival. First, we must address the understandable wish to avoid the pain and mess of politics in fashioning a vision of a common future. We argue that there is no other way, but why do we think so? As reviewed in chapter 3, the evidence—the fictional accounts of utopias and the attempts to build actual alternative societies in the real world—suggests that such a wish is no more than a sentimental delusion. Before moving on to that discussion, we clear the way by briefly reviewing those requirements of sustainability that are technical in nature—the guidelines that should be the ground rules for a politics of sustainability.

Notes

1. Anonymous, "China diverts Yangtze for dam construction," *Civil Engineering*, January 1998, 11.

2. A 150-watt light bulb (or two 75-watt light bulbs) left on for 10 hours consumes 1.5 kilowatt-hours (kWh) of electricity. Carbon emissions from using electric lights depend on where the electricity comes from (a hydroelectric,

coal-fired, natural-gas–fired, or nuclear generating plant, for example), but the average carbon emission rate for electricity delivered to residences from coal-fired plants in the United States is about 210 pounds of carbon dioxide per million British thermal units (Btu). One kWh equals 3,412 Btu, so 1.5 kWh equals 5,118 Btu. Dividing 5,118 Btu by 1 million Btu and multiplying the result by 210 pounds of carbon dioxide per million Btu yields 1.1 pounds of carbon dioxide. Conversion factors are from Energy Information Administration, *Annual Energy Review 1997*, tables B1 and C1.

3. Debora MacKenzie, "The turn of the worm," *New Scientist*, January 10, 1998, 13.

4. The figure is about 25 percent, expressed as net primary productivity, according to one celebrated estimate. See Peter Vitousek et al., "Human appropriation of the products of photosynthesis," *BioScience* 36(6) (1986): 368–373.

5. See Rick Potts, *Humanity's Descent: The Consequences of Ecological Instability* (New York: William Morrow, 1996).

6. Ibid., 68.

7. David Pearce et al., *Blueprint 3: Measuring Sustainable Development* (London: Earthscan Publications, 1993), 3.

8. Richard B. Norgaard, *Development Betrayed: The End of Progress and a Coevolutionary Revisioning of the Future* (London: Routledge, 1994), 11.

9. Larry Rasmussen, "Toward an ethics of sustainability," in Trent Schroyer, ed., *A World That Works: Building Blocks for a Just and Sustainable Society* (New York: Bootstrap Press, 1997), 352.

10. U.S. Secretary of State James Baker, speech to the National Governors' Association, February 1990, quoted in B. Rodes and R. Odell, *A Dictionary of Environmental Quotations* (New York: Simon and Schuster, 1992), 271.

11. Robert Repetto, *EPA Journal* (July–August 1990), quoted in Rodes and Odell, *Dictionary of Environmental Quotations*, 271.

12. World Commission on Environment and Development, *Our Common Future*, 1987, various sources.

13. Thomas M. Power, *Lost Landscapes and Failed Economies: The Search for a Value of Place* (Washington, DC: Island Press, 1996), 23.

14. An assessment of remaining global oil reserves by two experienced oil experts suggests that world production of oil from conventional wells (as opposed to shale, tar sands, etc.) will peak sometime in the next ten years,

unless a major recession restrains demand. See Colin J. Campbell and Jean H. Laherrère, "The end of cheap oil," *Scientific American,* March 1998, 78–83.

15. A few business leaders, some of them prominent, view global warming as a business opportunity. British Petroleum CEO John Browne, for example, reportedly believes that renewable energy could meet half the world's needs in fifty years and is driving his company to be a leader in the field. BP now controls about 10 percent of the global solar energy market. See Catherine Arnst, "When green begets green," *Business Week,* November 10, 1997, 98.

16. See Joseph Tainter, *The Collapse of Complex Societies* (Cambridge: Cambridge University Press, 1988).

17. Formally defined as "politics . . . where conflict is resolved in the absence of an independent ground through a participatory process of ongoing, proximate self-legislation and the creation of a political community capable of transforming dependent, private individuals into free citizens and partial and private interests into public goods" (Benjamin R. Barber, *Strong Democracy: Participatory Politics for a New Age* [Berkeley: University of California Press, 1984], 133). Among other things, that means greater emphasis on self-government by citizens rather than near-total reliance on government by elected representatives.

18. Barber, *Strong Democracy,* 258.

19. Ibid., 133.

Chapter 2
Minimum Technical Requirements for Sustainability

"What passes as definitions of sustainability are . . . often predictions of actions taken today that one hopes will lead to sustainability."[1] In other words, guesses. Nobody really knows what sustainability's minimum technical requirements are. The resources we need from the environment depend heavily on how many of us there are and how much each of us consumes. A high global population aspiring to a high standard of living will require much more of the available resources than a low population living more modestly. Only our children or grandchildren will know what the requirements are, or might have been, when the effects of current trends in population and consumption have become fully clear.

On the other hand, simply waiting for absolute clarity is dangerous. Given the signs of ecosystem degradation that have resulted from current levels of population and economic activity, and the skyward trajectory of global economic and population growth, it is reasonable to be worried. The human presence on Earth is probably not sustainable in its present size, habits, and tendencies.[2] Making an educated attempt at describing the necessary conditions for sustainability would be prudent, bearing in mind that ideas will change as new knowledge emerges and as values evolve.

In one sense, the minimum technical requirements for sustainability are very simple. The global ecosystem does three things that the human economy cannot do without, or do for itself. First, it provides resources (food, fiber, fuel, biological diversity, drugs, etc.). Second, it performs ecological services, such as photosynthesis, atmospheric gas regulation, climate and water regulation, soil formation, and pest control (see sidebar 2.1). Third, it absorbs wastes.[3] Running a sustainable economy therefore boils down to three simple rules: (1) Don't use up all the resources; (2) Don't undermine the delivery of ecological services; and (3) Don't overwhelm the waste-absorption capacity. These rules will be elaborated below. The first task is to lay a foundation.

Sidebar 2.1. Ecosystems at Your Service

Among the obvious benefits of the natural environment are the lumber, gravel, iron ore, water, and thousands of other valuable commodities we extract from it, but few people stop to think about how indispensable are the *services* the global ecosystem provides.[4] Without the immense diversity of the living things that constitute the ecology of Earth, there would be no human economy. The planet simply would not work.

The entire food web, for example, including the plants and animals humans consume, rests on a foundation of primary producers, which are the plants and sea-dwelling phytoplankton that use the energy of sunlight to make biomass through photosynthesis. Directly or indirectly, almost all creatures that are incapable of photosynthesis depend on this output. Photosynthesis is also the process by which the atmosphere, over billions of years, has come to be laden with oxygen, which humans and other aerobic creatures need for metabolism. Oxygen, in turn, forms the basis of the stratospheric ozone layer, which protects life on the ground from the harmful effects of the sun's ultraviolet radiation. Ecosystems seem generally to regulate the gaseous composition of the atmosphere.

Ecosystem services also include regulation of climate and the hydrological cycle. Plants draw water up through their roots and pump it into the air through transpiration, whence it returns to the soil as precipitation. While in the air, water vapor is by far the most significant greenhouse gas, thus contributing to the temperatures that make Earth a congenial place to live. The cycles of evaporation, condensation, and precipitation, to which plants contribute, affect the character of regional climates. By anchoring soil and breaking the force of falling rain, plants also help control flooding and soil erosion. Soil is both an ecosystem product and an ecosystem. Plants help make it by accelerating the weathering of rocks. Soil isn't just rock powder, though; each cubic yard contains millions of insects and billions of bacteria and other microorganisms that promote plant growth and decomposition.

Finally, pests would be far more serious competitors for our food were it not for the ecosystem services that help keep most disease organisms and troublesome insects in check. Many crop species rely on natural pollination, including wild bees, to maintain fertility.

How much is all this worth? In one sense, these services are priceless, since the world we live in would grind to a halt without them. One serious attempt to monetize global ecosystem services estimates their value at $33 *trillion* a year[5]—roughly twice the total output of all the world's national economies.

The Way We Wear

Conventional (neoclassical) economics is the reigning economic worldview of our time. Its assumptions are taught in every basic college economics course and reflected in every economics story in the newspaper. Its precepts guide the relations of governments with business and with each other. They are so pervasive as to be almost undetectable, like water is to a fish. The central assumption is that the workings of the economy are fundamentally captured by the circular flow of exchange value: firms produce goods, which are bought and consumed by households, which in turn supply the labor by which the firms produce the goods, and so on, as shown in figure 2.1.

The flow is energized by inputs of labor, capital, and land (i.e., natural

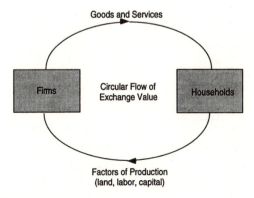

Figure 2.1. The neoclassical model of the economy.

resources, natural capital), and the transactions take place in markets. These markets, as described by conventional economic theory, are surprisingly abstract: they are "merely the meeting places of producers and consumers stripped of all history, social context, and biophysical reality. Place is reduced to transportation costs, and time to a single point—the immediate present."[6] Because the system is considered to be essentially self-contained, there is no reason it cannot keep growing indefinitely.

Conventional economics assumes that the economy and the ecosphere basically have little connection with each other, and land as a factor of production is relatively unimportant. This is in part because resources are considered to be almost infinitely substitutable for one another. If you run out of a resource, or if it becomes scarce and therefore too expensive to use, you can always switch to something else that is more abundant and cheaper. Also, the factors of production are treated as strongly substitutable. In agriculture, for instance, topsoil loss and compaction is compensated by more intensive use of fertilizers.

Contrast these notions with those of ecological economics, which argues that not only is land (the global ecosystem) more than a trivial factor in production, it is the economy's home and workshop, the very ground of its being. The economy nests within the global ecosystem and is utterly dependent on it. Further, the ecosystem is (1) limited in size, (2) not growing, and (3) not receiving any new flows of materials (though it is, fortunately, receiving a continuous flow of energy from the sun), as shown in figure 2.2.

Because economic production is basically the process of converting

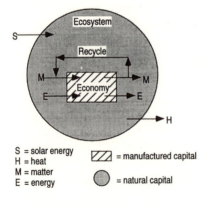

S = solar energy
H = heat
M = matter
E = energy

⬡⬡⬡ = manufactured capital

⬤ = natural capital

Figure 2.2. The economy as an open subsystem of the global ecosystem.

the natural world (renewable and nonrenewable resources and the ecosystems they constitute) to the manufactured world (houses, cars, computers, roads, books, plastic toys, etc., and nonnatural ecosystems such as parks and fields), the economy can grow only at the expense of the global ecosystem. Things tend to wear out, so they must be maintained and eventually replaced when they are worn beyond repair, through inputs of more resources. To increase economic growth so more people can have more stuff requires increasing the throughput of resources even more. To build more houses, for instance, requires clearing more forests to make room to site the houses and to make the lumber for their construction. Since no subsystem can outgrow its host, the economy cannot grow larger than the global ecosystem. The ecosystem is finite because the planet is not getting any bigger. This leads inescapably to the most important idea in ecological economics: *Economic growth cannot continue indefinitely.*

In the ecological economics view, resources flow into the economy from the enfolding ecosystem, are transformed by labor and capital (using energy, also a resource), and then pass out of the economy and back into the ecosystem in the form of wastes.[7] Many important transactions, such as the imposition of the costly health effects of smog, take place largely outside of markets.

Ecological economics argues that the factors of production ultimately are not substitutes but *complements.* They are substitutable to a certain extent: a bank teller (labor) can be replaced by an automated teller machine (capital), for example. Likewise, if you take away a logger's ax and give him a chain saw instead, he will be able to do the work of several men. Replacing all the loggers with chain saws, though, would not increase the output of trees but reduce it to zero. All forms of capital are necessary for economic production. Beyond certain critical limits, substituting one form of capital for another reduces output.

This is an important point, because it helps make the case for conserving ecosystem resources. To understand this fully, we have to answer the question, What actually do consumers consume?

Physics teaches that it is not matter or energy. These are neither created nor destroyed, only rearranged. Economic production, as the process of turning natural capital into manufactured capital and wastes, is the rearrangement of matter and energy. Felling a tree to make lumber for

furniture is an act of economic production that uses energy to reconfig-
ure the matter in the tree and thereby add value to it. But why are chairs
made from trees and not from dirt instead? (Dirt would be a lot cheaper,
after all.) Because the wood in the tree already possesses structural and
aesthetic qualities—strength, lightness, easy workability, beauty—that
make it a superior material for chair-making. That is, the tree *already has
value that was created by nature,* not added by economic activity.[8] Likewise, oil is useful in part because nature has spent eons applying geological forces to make it and concentrate it into deposits that can be tapped
relatively easily. When the chair wears out and is junked, or the oil is
extracted, refined, and burned, both the value inherent in the resource
and the value added during the chair's manufacture or the oil's refining
have been consumed. To replace them will take more economic production—and more work by nature to make another tree or (in a few hundred million years) more oil.

The economy is not really the closed circular flow so central to conventional economics but rather a constant one-way flow of high-quality
matter/energy that is turned into low-quality matter/energy (wastes).
Economic activity is *entropic;* ultimately, it increases disorder. We take coal
and trees and turn them into carbon dioxide, waste heat, and landfill
mass. (They only pause along the way in the "chair" phase.) The laws of
thermodynamics say that the total quantity of matter/energy is the same
before and after, but the value or usefulness of the stuff has declined
sharply. So what consumers actually consume is the *value* in stuff—that is,
the value of both natural and manmade capital.

Converting natural capital to manufactured capital adds value of one
kind but subtracts value of another kind. Value is added to the felled tree
as it is turned into lumber, but value is also subtracted because the tree's
contribution to the services of the forest ecosystem—habitat, water regulation, oxygenation, evapotranspiration, and so on—is lost even as the
services of the lumber are gained. (Measures of economic output, such as
gross national product [GNP], typically ignore the loss of value from cut-
ting the tree.)

A big economy means high rates of conversion of natural capital to
manufactured capital and high throughput of resources into the economy
and out again as wastes. Although we think of economic activity as cre-

ative, it is destructive, too. Humans get to use the created stuff while it's passing through, but everything eventually wears out and has to be replaced, which means drawing on stocks of resources and energy. With nonrenewable resources, the bigger the economy grows and the more throughput there is, the faster they approach depletion. Renewable resources can replenish themselves, but only if the rate at which they are used up does not exceed their regeneration rate. The record of depletion of rain forests and fisheries, among other renewables, suggests that the global economy is already harvesting many of them faster than they regenerate (see sidebar 2.2). As the amount of manufactured capital increases and the amount of natural capital dwindles, a point eventually is reached at which the natural capital sacrificed to make manufactured capital is worth more (as biodiversity, ecosystem services, etc.) than the stuff (consumer goods, magazines, or whatever) that is made from it. More value is lost by drawing down stocks of natural capital than is gained by converting it to build up stocks of manufactured capital. At that point—which we may already have reached—further economic growth actually makes people worse off in the long run, not better. GNP contin- ues to rise, but the widely ignored ecological side of the ledger reveals shocking losses.

To return to our three rules about resources, ecological services, and wastes, we can now see more specifically why endless economic growth is not sustainable. Any economy, growing or not, uses resources. Some resources, such as oil and minerals, are finite in quantity and nonrenew- able. Natural processes concentrate them into useful deposits at geologi- cal rates that are of no help to humans. Further, most of the best deposits have already been tapped. Concentrations of copper ore mined in the United States, for example, have been falling for decades. The output of every oil field follows a predictable pattern (called the Hubbert curve after the geologist who discovered it): rapidly rising production to a peak, at which point about half the oil is gone, then an almost equally rapid decline. Eventually, the time will come when the richest and most accessi- ble supplies will be exhausted and the remainder will be too expensive to extract and use.

Technological optimists argue that new technologies, evoked by the marketplace's response to scarcity, will enable the economy simply to tap

Sidebar 2.2. Allowing for Depletion

Warnings of resource scarcity are not new, and they have been proven wrong often enough to engender skepticism. The evidence keeps mounting, though, to indicate that we are using up many resources too fast and without adequate thought for the future.

The world's ocean fisheries, for instance, appear to be in serious trouble, according to a recent report by the United Nations Food and Agriculture Organization:

> About 35% of the 200 major fishery resources are senescent (i.e., showing declining yields), about 25% are mature (i.e., plateauing at a high exploitation level), 40% are still 'developing,' and 0% remain at low exploitation (undeveloped) level. This indicates that around 60% of the major world fish resources are either mature or senescent, and given that few countries have established effective control of fishing capacity, these resources are in urgent need of management action to halt the increase in fishing capacity or to rehabilitate damaged resources.[9]

Another example is oil. The fear of running out of it is a perennial favorite worry. That prospect seems absurd at the time of this writing (1999), with world markets awash in crude oil. Yet forecasts of the end of cheap oil are common—and now they are being made by geologists, energy experts, and oil company executives. The head of the Italian oil company ENI believes that world oil production will peak by 2005 and then begin an inexorable decline, even as demand continues to rise.[10] Two petroleum geologists with more than eighty years' oil industry experience between them recently argued that "there is only so much crude oil in the world, and the industry has found about 90 percent of it." Unless there is a global recession, they predict, world production of conventional oil (as opposed to oil from shale or tar sands) will probably peak by 2010.[11] Major oil companies, such as British Petroleum and Royal Dutch Shell, have seen the handwriting on the wall and are moving into the renewable energy business.[12]

some other resource stocks. That could happen, but there is nothing simple or predictable about it. The effects of technological "progress" on economic output are somewhat ambiguous, since it cannot be measured directly. Much of what appears to be technology-driven increases in output may actually be from increases in resource throughput.[13] Technologies frequently also have what historian Edward Tenner calls revenge effects. Hard-shell football helmets and other armoring, for example, make football more dangerous, not less, leading to higher injury rates.[14] Automobiles seem to promise the ultimate in freedom and mobility—until everybody gets one and the roads become gridlocked. Technologies sometimes simply have nasty and unexpected side effects. In the environmental arena, the classic example of this problem is chlorofluorocarbons (CFCs), which appeared miraculously beneficial when they were introduced, in 1935. Only decades later was their disastrous effect on stratospheric ozone discovered. In short, while technology keeps changing, to argue that the changes always and predictably amount to progress and unmixed blessings—and that they allow us therefore to escape resource constraints—is naïve.

As for renewable resources, the economy can draw on them forever so long as its rate of use does not exceed their regeneration rate. Sunlight is the engine that drives Earth's natural systems, and as long as the sun shines, the stocks of fish, trees, and other renewables will strive to replenish themselves. And they will succeed, unless economic activity interferes with the conditions necessary for their survival. Leaving certain minimum stocks undisturbed allows a perpetual flow of "interest income" for use in the human economy. This interest income is proportional to the size of the stocks. Like a bank account, drawing down the principal reduces the flow of interest, and spending all the principal stops the interest flow for good. Some stocks of renewable resources, such as the fisheries mentioned earlier, may already be ravaged beyond the point of recovery.

Though we speak of them as stocks, renewable resources are not just passive assets, like so many gold bars piled up in the vaults at Fort Knox. As noted earlier, they are dynamic parts of a vast and complex global ecosystem that provides essential services to the economy (see sidebar 2.3 for an example of how valuable these services can be in monetary terms). Tropical rain forests, for instance, regulate water cycles and soil temperatures, control flooding, store biodiversity, sequester carbon through photosynthesis, and so on. Clearing rain forests wipes out species, raises soil temperatures, releases carbon into the atmosphere, and disrupts local rainfall patterns, making the ecosystem less supportive of future economic

---

Sidebar 2.3. The Price of Clean Water

New York City has learned what natural ecosystem services are worth: for the city's water supply, about $4.5 billion. That's how much money may be saved by restoring the Catskill watershed areas that supply New York City with tap water, rather than building a huge filtration plant.[15]

The soils and vegetation of the Catskill watersheds, 125 miles north-west of the city, are natural filtration systems. As development proceeds, their effectiveness is compromised and so is the quality of the water in the six Catskill lakes that supply New York with 90 percent of its water. Declining water quality and new federal regulations had brought the city face-to-face with the prospect of spending up to $6 billion on a filtration plant to clean the waters collected in the watershed and piped downstate to New Yorkers' faucets.

New York and thirty Catskill communities in the watershed chose a different route. In 1997, after arduous negotiations, they signed the New York City Watershed Agreement, a $1.4 billion plan to preserve the watershed's natural filtration and purification capacity. The plan will allocate $550 million to control development and preserve the farms and woodlots of the watershed, and another $660 million to repair aging sewage treatment facilities and fix old septic systems. Additional funds are being spent on programs to teach dairy farmers and woodlot owners how best to manage their properties and reduce runoff that pollutes waterways.

The result is a plan that not only is less costly but protects important social values as well. Residents in the upstate communities find that the lands they manage have value beyond those assigned by the traditional marketplace—value that, once acknowledged, can help preserve traditional ways of life in the country. City dwellers not only get clean water at less cost, they benefit from the preservation of green space and rural amenities around the city. The agreement embodies the dawning recognition of how city and country need each other, of how, as was said about another community in the West, "Their fates were woven together, how the good life that they each wanted depended upon the other's being secured in a different but equally good life."[16]

---

activity in a variety of ways. Economic overexploitation can thus compromise dozens of essential ecological services and undermine the economy in the long run even as it appears to stoke the economy in the short run.

Finally, there is the matter of waste absorption, the least-appreciated function of the ecosystem. As an entropic activity, the economy inevitably produces wastes and pollution. Even heroic recycling efforts cannot entirely prevent the production of waste. It has to go somewhere, whether it is in the form of fluids, solids, gases, or heat. That "somewhere" is the ecosystem, which is the "sink" of last resort.

The ecosystem has a remarkable capacity to absorb, transform, and render harmless many wastes. But it can be swamped by asking it to do too much or by presenting it with waste substances it has no mechanisms to handle. An example of the former is the buildup of carbon dioxide in the atmosphere due to major additions of the gas from human sources, mainly fossil fuel combustion and deforestation. Not all the carbon dioxide produced from these sources stays in the atmosphere; much of it is absorbed by means that are not fully understood. The steadily rising concentrations, however, strongly suggest that even nature's unexpected abilities to cope with carbon dioxide are being overtaxed.

The problem of alien substances is of a different order. The sophisticated waste-absorption mechanisms built into the ecosystem involve millions of plant and animal creatures and thousands of processes working in intricate harmony. They evolved over millions or billions of years to handle essentially any category of waste the ecosystem could generate; every organism's waste was some other organism's fuel. But only in the last few decades, humans have learned how to make substances (and even elements) not found in nature. We have also begun to use substances, such as radioactive materials, that are imported in unnaturally large quantities from Earth's crust into the living communities of its ecosystem. Many of the new or imported substances are known to have, or may prove to have, undesirable effects on living creatures. Some are outright poisons. They often accumulate in animal tissues and increase in concentration at higher levels of the food chain. The ecosystem does not necessarily have any mechanisms to render these substances harmless because there has not been enough time for such mechanisms to evolve.

In at least two ways, then, the waste absorption capacity of the ecosystem can be overwhelmed. Because a system's wastes are by definition what it rejects as useless or harmful to it, allowing unassimilable wastes to build up could be dangerous or lethal to human well-being or economic activity (which is supposed to serve human well-being). Examples of the consequences, new and old, local and global, are plentiful. In medieval London, the air was so polluted by smoke from thousands of coal-fired hearths that in 1306, King Edward I banned the use of coal except by blacksmiths. (The ban was ignored because fuelwood was scarce enough by then that coal was economically necessary.) Industrial smokestacks of the nineteenth and twentieth centuries belched noxious gases, ash, and particulates that proved to be costly and unhealthful to the people living downwind. (Initial response: raise the smokestacks to spread their output over a greater area—a classic application of the principle that "the solution to pollution is dilution.") The piles of waste rock and soil from mining operations, called tailings, often contain potent chemical compounds that can be leached out by rainfall into nearby watercourses, acidifying or otherwise altering the chemistry of streams in ways that harm fish and other aquatic creatures. Tailpipe emissions from cars and trucks interact with sunlight to create smog, which causes respiratory problems, especially in asthmatics and the elderly. A recent development that shows how pervasively wastes can migrate through the environment is the buildup of excreted pharmaceutical drugs and their metabolites in surface waters, groundwater, treated tap water, and even the oceans.[17] The effects are little studied as yet.

Earth is often compared metaphorically to a household, but there are some pertinent differences. A household imports the resources it needs from somewhere else. A household's services are bought as commodities and, so long as the bills are paid, are not affected by how it is run. A household can throw its trash and garbage away, and so avoid fouling itself. But a household is an open system, and Earth essentially is a closed one. At the global scale, there is no "somewhere else" or "away," and the abuse of services undermines their delivery. Can a globalizing economy be run in a way that prevents it from destroying the ecosphere that makes it possible?

House Rules

The three rules mentioned at the beginning of the chapter—conserve resources, protect ecological services, conserve waste-absorption capacity—are easy to state but difficult to follow. The economy offers countless immediate payoffs for ignoring the rules and few for observing them. In the United States, gasoline costs less than bottled water, so why *not* buy a 2-ton, four-wheel-drive sport utility vehicle for running around the suburbs? Mass-production technology and low wages in overseas labor markets make an electric mixer cheaper to replace than fix, even if repair did not require shipment to an "authorized service center" hundreds of miles away.

Deluged by advertising and conditioned to equating happiness with acquisition, people buy, buy, buy. Pervaded by a culture of expansion, enrichment, and predatory competition, corporations pursue growth for its own sake. Wedded to a vision of prosperity that intrinsically favors consumption, politicians tremble when quarterly GNP increases slip. It is no surprise that the economy's enormous momentum is in the wrong direction. However, the economy could be restructured so that people are given incentives to act in more sustainable ways. Taxing consumption and resource use rather than income, for example, would send a more accurate price signal about the value of resources and thus encourage more careful use of them. There are no technical reasons why this and a host of other useful reforms could not be carried out, only political ones.

Of course, the obstacles to change can be entrenched and formidable, whether one is a local activist or a reform-minded member of the ruling class. The speed-hump story recounted in chapter 1 was a modest but vivid personal lesson in the power of the status quo at the neighborhood level. At a higher level, former secretary of labor Robert Reich tells of a meeting with other Clinton administration officials at which he proposed ending certain forms of corporate welfare:

> At today's budget meeting I suggest to them we finance the education and training tax break by closing some tax loopholes. In deference to Bentsen, I avoid any mention of oil and gas. What about the tax breaks for the insurance industry?
>
> BENTSEN: That would be *very* unwise, politically.
>
> STEPHANOPOULOS: Republicans would accuse us of raising taxes.
>
> PANETTA: The insurers would be on top of us.

RUBIN: The financial markets would take it badly.

BENTSEN: Don't even *think* about it.

I gingerly offer up another one: the advertising industry claims that advertising builds up a company's goodwill with customers for years, right? So take them at their word. Prevent companies from deducting the entire cost of their advertising right away. Make them treat advertising like any other investment and deduct its cost over several years. This would save the Treasury billions.

PANETTA: *Advertisers?* Are you *kidding?* We'd have the media all over us.

BENTSEN: A nonstarter.

STEPHANOPOULOS: Forget that one. And don't repeat it outside this room. [Laughter]

RUBIN: The financial markets would take it very badly.[18]

Resistance to change abounds at all levels. Nevertheless, things do change. The only questions are who will drive the changes and in what direction? At the heart of this book is the belief that political obstacles to reshaping the economy can be overcome, finessed, or subverted. So let's stipulate, for the moment, that the economic system can be refashioned for sustainability. Such a system like this would do a few things differently:[19]

PAY ATTENTION TO SCALE. First and foremost, it would both measure and control the scale of the economy relative to the global ecosystem. The need to control the economy's size so that it does not consume its host ecosystem should already be clear. Obviously, it would be useful to have measures applicable to both so they could be readily compared.

At present, the size of any economy is normally expressed in terms of gross national product (GNP), which is basically a measure of how fast money changes hands (more formally, it is the dollar value of all consumption goods and services, government spending, investments, and net exports). Economists know that GNP is flawed in many ways. For instance, GNP is widely used as a proxy for national well-being, yet it makes no distinction between the money paid to a music shop for a new grand piano and the equivalent sum charged by a physical therapist to retrain an auto-accident victim to use her limbs—which doesn't so much add to well-being as try to restore its loss. By the GNP measure, the more auto accidents, the better off the nation is. Also, the GNP does not count unsold output, such as child care and the household labor performed by parents.

With respect to the scale problem, GNP has several shortcomings. One is that it uses units of value (e.g., dollars) and so counts both growth in physical throughput (more stuff flowing through the economy) and increases in the value of goods or services, whether or not they require more stuff. Another is that many of the things that need measuring are not captured in markets and so are not directly monetizable. Attempts to attach dollar values to ecological goods and services are controversial, but the study we mentioned earlier valued global ecosystem services at a central value of $33 trillion a year, with a range of $16 trillion to $54 trillion per year, or roughly one to three times global GNP. That obviously reinforces the conviction that ecosystem assets are valuable but might suggest a dangerous train of thought: if natural capital is three times larger in dollar terms than global GNP, then there must be plenty of room for the economy to grow. Why worry yet? But this imputation of a dollar value to natural capital is a measure not of its size but of its indispensability. Compare two worlds: In one, the combined dollar value of the economy and the ecosystem is split in a ratio of one to three. In the second world, the total value is the same but ratio is reversed, so the global economy is valued at three times the remaining natural capital. Are the two worlds equivalent? Of course not: the second one is radically poorer in the assets it needs most. Like the chain saw–equipped logger without a forest, without natural capital there is no economy. Many consider natural capital priceless—because we literally cannot do without it—and therefore of infinite value. It hardly follows that we can never use it up.

More useful measures of scale are needed. One possibility is the economy's appropriation of net primary productivity (NPP). NPP is basically the output of all Earth's photosynthesizers, which include land-dwelling plants and marine phytoplankton. Primary producers turn sunlight into biomass and form the foundation of the food web. The rate of biomass output is finite because the rate of the flow of solar energy to Earth is essentially fixed. If the human economy takes over, captures, or displaces more and more of this output—by consuming more plants and animals, harvesting more trees, turning forests into less-productive pastureland, building houses, roads, parking lots, and skyscrapers, and so on—then less and less of the supporting ecosystem is left. One study estimates that by the mid-1980s, the human economy had appropriated about 40 percent of Earth's land-based NPP.[20] With continued economic growth, that figure will rise. Can the global economy survive over the long term if it appropriates 50 percent? 80 percent? No one knows.

RESTRAIN AFFLUENCE AND POPULATION. The scale of the global economy is mainly a function of the total number of people and their standard of living. Both greater numbers and greater per capita wealth mean more conversion of natural capital to built capital, so to control scale it is necessary to control total population and average affluence. Technology can mediate this effect (see below), but the limits imposed on economic growth by the finite size of the ecosystem mean there is no eliminating it. At some hypothetical sustainable maximum level of resource use, there is a direct trade-off between population and per capita wealth. This effectively caps the arithmetic product of people × wealth. Within that limit, many scenarios are possible. If it turns out that Earth can support, say, 15 billion desperately poor people without exceeding its carrying capacity, then perhaps it could support 3 billion people very comfortably and for an indefinite period. How the total wealth is distributed is ultimately a political problem, but clearly, the more people who inhabit the planet at one time, the worse the average person's chance for a decent life in material terms.

A sustainable per capita level of wealth is difficult to estimate, but it is very doubtful that 10 or 15 billion people all will be able to live like today's average American does. There is roughly three-fifths of a car for every person in the United States—a major reason why Americans make up about 5 percent of the world's population and use 25 percent of the energy—and those cars exact a steep ecological price in pollution, non-renewable resource depletion, and pavement-related loss of NPP. Does it seem prudent to assume Earth can support a car in every Chinese and Indian garage? (And if not, the Indians and Chinese might reasonably argue, why should Americans get all the cars? We have been talking about average per-capita wealth, but in fact human society is marked by vast differences in wealth between the rich and the poor. The fact that Earth can tolerate only a certain fixed limit of wealth generation from its stocks of natural capital raises the very difficult-but inescapable-question of redistribution).

ACKNOWLEDGE THE INHERENT UNCERTAINTY OF THE ECONOMY AND THE ECO-SYSTEM. Human economic activity is changing the natural world. Certain major trends are undeniable, including rising atmospheric concentrations of carbon dioxide, human modification of the global nitrogen cycle (mainly by means of fertilizer use and tailpipe emissions), and human alterations of land use and cover.[21]

Still, we float on a sea of uncertainty. Sometimes we are swamped by it. The world is an economically, politically, culturally, and ecologically complex, even chaotic, place. It has many players and forces, all interacting with each other and changing over time. Changes in one player or force influence the state and evolution of the others. Such a complex, coevolving system inevitably and endlessly produces surprises, both major and minor, creating a vexing element of inherent unpredictability that cannot be eliminated no matter how much we learn about the system.[22]

Under these circumstances, nothing can be decided once and for all. Decisions, tendencies, policies, regulations, and institutions always will be made obsolete by unexpected events, forcing us to reconsider what we think we know and how, therefore, to change what we do. This book repeats the familiar argument that the world's cultural and political commitment to endless economic growth undermines itself by straining the support capacities of the global ecosystem. Humanity probably is in for some serious adapting over the next few decades. None of this is new, of course. What has been missing is a way to routinely change decisions and policies when events make it unmistakably clear that they are no longer working. Humans are among the most adaptive species—we can usually figure out how to make do and get by. Facing the point of an ecological knife, no doubt we will muddle through. It would be better to adapt, though, learning to efficiently and systematically test and reevaluate the policies and institutions that define our relationship with the biosphere.

What's needed is a way to institutionalize adaptability. A key to this process is adaptive management, a concept developed by a team led by ecologist C. S. Holling and described here by political scientist Kai Lee:

> [Adaptive management is] an approach to natural resource policy that embodies a simple imperative: policies are experiments; learn from them. In order to live, we use the resources of the world, but we do not understand nature well enough to know how to live harmoniously within environmental limits. Adaptive management takes that uncertainty seriously, treating human interventions in natural systems as experimental probes.[23]

It is often said that we are conducting a great experiment with the world, which is true. But it is not a very good experiment. A proper experiment would be a carefully designed and controlled inquiry, carried out by serious, thoughtful people who have struggled to free themselves of ideological assumptions, who appreciate the potential consequences, and who are eager to learn from whatever outcome it yields. What we're doing with the world is more like a child playing with matches, or an eighth-grade chemistry student randomly mixing reagents just to see what happens. The result might be Flubber, but it's more likely to be an explosion or a cloud of toxic gas.

To succeed, an experiment requires clearly defined expectations, precise measurement, and careful comparison of outcomes with expectations to create reliable new knowledge. One of adaptive management's benefits is that it teaches valuable lessons even when things don't turn out as expected—that is, when the ecosystem's inherent uncertainties produce their inevitable surprises.

Adaptive management is a key element of what Kai Lee calls civic science, the practice of science in a political setting where the science is used to inform, but not dominate, political action.[24] (We will further discuss science's role in pursuing sustainability in chapter 5.)

STRESS DEVELOPMENT, NOT GROWTH. Make the economy better, not bigger. The purpose of economic activity should not be to increase for its own sake but to improve human well-being. Growth has considerable, but finite, capacity to do that. Clearly, the very poor benefit enormously from growth that improves their material standard of living, and it is to the world's poor that the benefits of future economic growth ought mainly to be channeled. A 20 percent increase in income improves well-being a lot more for a Bangladeshi family earning $500 a year than for Bill Gates (net worth as of January 1999: roughly $80 billion).

But beyond a certain point, growth not only is uneconomic in ecological terms (more value is lost than gained in converting natural capital to built capital), it cannot significantly improve well-being, either, because much accumulation of personal wealth is driven by considerations of relative status in society.[25] People want things because others have them or, conversely, because others *don't* have them (yet). This tendency to seek status through possessions is often reviled, perhaps unfairly so, to the extent that it simply reflects the natural human wish to belong by having what others have.[26] But it cannot work forever; as the old saying goes, if everyone in the bleachers stands on tiptoe, nobody can see any better. Moreover, the anthropological evidence shows that consumerism isn't "natural" but learned, and that essential human needs and wants can easily be met outside the perpetual he-who-dies-with-the-most-toys-wins paradigm. Estimates of the value people attach to leisure and satisfying work suggest that their contribution to their well-being is far greater than that of the actual goods and services the economy produces.[27]

When Earth's carrying capacity is reached (preferably before), further increases in well-being can and should be achieved by development without growth. If too much growth threatens to crack the ecological foundation of the economy, development is the only viable option. Development is generally taken to mean technological improvement, though it would be a mistake to confine it to that definition. Technology alone won't be enough.

There is room for technology to improve in every sphere, and evidence of progress appears daily. Computers, carbon-fiber automobiles powered by fuel cells, and closed-cycle ("no-waste") manufacturing techniques and processes are all examples of doing more with less. After a long history of using technical improvements mainly to raise resource extraction efficiencies and thereby boost throughput levels, industry is beginning to explore the possibilities of "dematerialization" and "industrial ecology":

techniques to build lighter, stronger, and more efficiently, to pollute less, to use less energy per unit output, and to model industrial processes on natural ones so that one factory's "wastes" become another's feedstocks. These developments, along with the inevitable conversion to renewable resources, are welcome and necessary (provided that their inevitable back-biting tendencies are not forgotten). The nurturing of them is the special responsibility of the global North (broadly, Europe, Japan, and North America), which is richer in technical know-how and capital than the poorer South. Because the North owns a disproportionate share of the world's wealth (much of it more or less purloined[28]), it is a fair burden.

But however much technological advances improve resource-use efficiency, it would be foolish to assume that this will free humanity from the growth constraints imposed by population and scale. Per capita gross national product in the world's high-income countries was about $27,000 in 1995. In the world's low-income countries, it was $430,[29] a factor difference of about 58. Without increasing throughput, resource productivity would have to improve 58 times overnight to enable the hundreds of millions of poor people on Earth to live as richly as Americans. If global population doubles in a generation, that factor goes up to about 116. How realistic is it to bet on resource productivity improving 116-fold in 30 or 35 years? Or even 10- or 20-fold?

BE WARY OF THE RISKS AND COSTS OF INCREASING COMPLEXITY. Powerful forces in society and the economy, especially the ambitions of nation-states and the urge of transnational corporations seeking the last remaining markets not already saturated with consumer goods, seem hell-bent on globalization at whatever cost. Globalization may be the final step in a process of increasing complexification of culture that began with the founding of settled agricultural societies several thousand years ago and has been accelerating ever since. However, there is evidence that as cultures or civilizations become ever more complex, they become increasingly difficult to sustain.

The complexity of a society, according to archaeologist Joseph Tainter, means its "size . . . , the number and distinctiveness of its parts, the variety of specialized social roles and distinct social personalities that it incorporates, the number of distinct social personalities present, and the variety of mechanisms for organizing these into a coherent functioning whole."[30] Hunter-gatherer societies, in which humanity lived for nearly all its history and prehistory, have only a few social roles and very little formal organization. Modern societies, in contrast, have used energy subsidies and technological advances to raise agricultural output so dramatically that a tiny fraction of the populace is all that is required to produce enormous surpluses. This has enabled a staggering increase in role specialization. The U.S. *Dictionary of Occupational Titles* alone, for instance, lists about twenty thousand formal job types, and this kind of classification disregards thousands of informal or nonoccupational roles (father, mother, groupie, cheerleader, waterboy, gadfly, barfly, etc.). There may be a million or more social personalities in industrial societies.[31] Besides this differentiation of roles, complex societies also tend to be marked by hierarchy and, often, inequality.

Tainter argues that complexity doesn't just happen: it is a strategy employed to solve a problem or secure a benefit. Because people often dislike and reject complexity—for instance, yearning for the "good old days" usually implies a wish to be free of the hassles of modern life—there must be a payoff for pursuing complexity. The payoff is that it usually works; it achieves greater wealth, power, security, or whatever, for someone. But complexity requires an investment:

> More complex societies are costlier to maintain than simpler ones and require higher support levels per capita. A society that is more complex has more sub-groups and social roles, more networks among groups and individuals, more horizontal and vertical controls, higher flow of information, greater centralization of information, more specialization, and greater interdependence of parts. Increasing any of these dimensions requires biological, mechanical, or chemical energy. In the days before fossil fuel subsidies, increasing the complexity of a society usually meant that the majority of its population had to work harder.[32]

Like many investments, the returns to increasing complexity tend to diminish as the level of investment rises. Eventually, a point is reached where the gains are not worth the increasing costs. At that point, the culture is vulnerable to collapse, which is a "rapid transformation to a lower degree of complexity, typically involving significantly less energy consumption."[33] Tainter uses this analysis to explore and explain the expansion and collapse of the world's great civilizations.

The cheap and abundant coal, oil, and natural gas that drive industrial society have made it possible to increase complexity in every area: science, medicine, transportation, information processing, weaponry, and our personal lives (as consumers, a role that barely existed before the twentieth century). Today, the ecological price of such intensive energy use is becoming clearer, and expanding energy use to support even greater complexity, in response to demands for higher standards of living here and in the developing world, will be difficult. Energy can be freed up by discovering and implementing ways to make more efficient use of it, but technological innovation alone can carry us only so far. Energy constraints will make it tougher to pursue the strategy of more complexity in trying to build a sustainable world. This helps explain why so many proposals for restructuring society for sustainability involve decentralization, regionalism, simpler living, and a focus on the quality of life rather than the quantity of individual wealth each person owns.[34]

Such a simplification could happen by nature's fiat, if the global ecosystem is pushed beyond its supportive and waste-assimilative limits, or by human choice. Tainter believes that avoiding collapse is unlikely unless a profound ecological crisis evokes wide popular support in the industrial nations for a voluntary reduction in the average standard of living. Yet it is arguable that the consequences of a voluntary simplification in the industrialized world, far from being cataclysmic, would range from modest to trivial. For example, reducing U.S. per capita energy use 30 percent overnight would take us back to the usage level of . . . 1900? 1925? No, 1960—the year John F. Kennedy was elected president.[35] Seems like only yesterday. Was life so barbaric then?

In view of humanity's twelve-thousand-year history of solving problems through investments in greater complexity, the likelihood of voluntary simplification may seem remote. We argue in the following chapters, however, that a different politics not only might make the case for a kind of simplification attractive and compelling but also bring it closer to our grasp. Chapter 3 discusses why it is that politics, of one kind or another, is unavoidable in pursuing sustainability.

Notes

1. R. Costanza and B. Patten, "Defining and predicting sustainability," *Ecological Economics* 15(3) (December 1995): 194.

2. We take it for granted that, whatever form of sustainability our collective politics defines and pursues, humans will continue to live on Earth. For the record, there is a small movement (presumably self-limiting) that advocates the voluntary extinction of human beings as a means of making the world safe for zoocracy.

3. P. Ekins, "Towards an economics for environmental sustainability," in R. Costanza et al., *Getting Down to Earth: Practical Applications of Ecological Economics* (Washington, DC: Island Press, 1996), 129–152.

4. P. Ehrlich and A. Ehrlich, "The value of biodiversity," 1991 (manuscript).

5. R. Costanza et al. "The value of the world's ecosystem services and natural capital," *Nature* 387 (May 15, 1997): 253–260.

6. J. Gowdy and S. O'Hara, *Economic Theory for Environmentalists* (Delray Beach, FL: St. Lucie Press, 1995), 9.

7. Energy follows this path, too. For example, coal must be dug from the ground, transported to power plants, and crushed so it can be burned for generating electricity. By-products include sulfurous gases and carbon dioxide, but also unusable heat. Because of the practical and theoretical limitations of conventional electricity-generating equipment, only about one-third of the chemical energy content of the coal that is fed into the boilers is actually converted to electricity delivered to end-users. The rest is lost as waste heat to the surrounding environment and does no useful work.

8. H. Daly, "Consumption: Value added, physical transformation, and welfare," in R. Costanza et al., *Getting Down to Earth*, 49–59.

9. United Nations Food and Agriculture Organization. *Review of the State of World Fishery Resources: Marine Fisheries*, FAO Fisheries Circular No. 920 FIRM/C920, Rome, 1997.

10. H. Banks, "Cheap oil: Enjoy it while it lasts," *Forbes*, June 15, 1998, pp. 84, 86.

11. C. Campbell and J. Laherrère, "The end of cheap oil," *Scientific American*, March 1998, 78–83.

12. G. Easterbrook, "The coming oil crisis—really," *Los Angeles Times,* June 7, 1998, M1.

13. H. Daly, *Steady-State Economics* (Washington, DC: Island Press, 1991), 106.

14. E. Tenner, *Why Things Bite Back: Technology and the Revenge of Unintended Consequences* (New York: Knopf, 1996), 217 ff.

15. E. Nickens, "Watershed paradox: New York City's water quality protection efforts," *New York Times,* January 1, 1998, sec. 4, 21.

16. D. Kemmis, *Community and the Politics of Place* (Norman: University of Oklahoma Press, 1990), 66.

17. P. Montague, "Drugs in the water," in *Rachel's Environment and Health Weekly,* issue 614, September 3, 1998.

18. R. Reich, *Locked in the Cabinet* (New York: Knopf, 1997), 213–214.

19. This section offers only a very brief introduction to these points, which are covered extensively in the ecological economics literature. General readers interested in more detail might consult *Natural Capital and Human Economic Survival,* 2nd ed. (Lewis Publishers, 1999). A more advanced volume is *An Introduction to Ecological Economics* (1997), from St. Lucie Press.

20. P. Vitousek, P. Ehrlich, A. Ehrlich, and P. Matson, "Human appropriation of the products of photosynthesis," *BioScience* 36(6): 368–373.

21. P. Vitousek, "Beyond global warming: Ecology and global change," *Ecology* 75(7) (1994).

22. This is possibly because, as Terry Pratchett's fictional philosopher Ly Tin Wheedle has it, "Chaos is found in greatest abundance wherever order is being sought. It always defeats order, because it is better organized" (*Interesting Times* [New York: HarperPrism, 1994]).

23. K. Lee, *Compass and Gyroscope: Integrating Science and Politics for the Environment* (Washington, DC: Island Press, 1993), 9.

24. Ibid., 161 ff.

25. See, for example, Richard Easterlin's classic 1974 study, "Does economic growth improve the human lot? Some empirical evidence," in P. David and M. Reder, eds., *Nations and Households in Economic Growth: Essays in Honor of Moses Abramovitz* (New York: Academic Press, 1974). Easterlin concluded that "economic growth does not raise society to some ultimate state of plenty. Rather, the economic growth process itself engenders ever-growing wants that lead it ever onward" (p. 121).

26. T. Scitovsky, *The Joyless Economy: The Psychology of Human Satisfaction* (New York: Oxford University Press, 1992), 115.

27. See Scitovsky, *The Joyless Economy*, 102–103, and J. Gowdy, ed., *Limited Wants and Unlimited Means: A Reader on Hunter-Gatherer Economics and the Environment* (Washington, DC: Island Press, 1998).

28. See, among countless other sources, C. Ponting, *A Green History of the Earth: The Environment and the Collapse of Great Civilizations* (New York: Penguin Books, 1991).

29. World Resources Institute, *World Resources, 1998–1999: A Guide to the Global Environment* (New York: Oxford University Press, 1998), table 6.1.

30. J. Tainter, *The Collapse of Complex Societies* (Cambridge: Cambridge University Press, 1988), 23.

31. Ibid.

32. J. Tainter, "Complexity, problem solving, and sustainable societies," in R. Costanza et al., *Getting Down to Earth: Practical Applications of Ecological Economics* (Washington, DC: Island Press, 1996), 63.

33. Ibid., 74.

34. For critical synopses of dozens of such proposals, see T. Williamson, *What Comes Next? Proposals for a Different Society* (Washington, DC: National Center for Economic and Security Alternatives, 1998).

35. See Energy Information Administration, *Annual Energy Review* (Washington, DC: U.S. Department of Energy, 1998), table 1.5. This projection does not even allow for the fact that we have been backsliding in energy efficiency for fifteen years. U.S. per capita energy use in 1997 was 352 million Btu, exactly the same as in 1978. In between, energy efficiency improvements (mainly driven by rising oil prices) reduced per capita use to 308.

Chapter 3
Aiming for Genotopia

The rational process of figuring out how to achieve a sustainable world must begin with a nonrational act of imagination. As Donella Meadows has pointed out, most discussions of sustainability focus on implementation and ignore the critical questions of what the world of our dreams would actually look, feel, and smell like.[1] The trouble is, the sustainable world generally offered by environmentalists is based on "restriction, prohibition, regulation and sacrifice. . . . Hardly anyone seems to envision a sustainable world that would be nice to live in."[2] This is a self-defeating lapse of imagination that could dim the prospects for achieving sustainability. There seem to be only two visions on the table. In the conventional vision, the human economy and population keep growing vigorously, and everyone eagerly chases the dream of greater consumption. The environmentalist point of view rightly denies the workability of this vision but offers in its place a kind of lifelong global celery diet. It is hardly surprising that most people choose the first path.

Visions are important because of their emotional and intellectual power. Two sharply contrasting visions of the near future, both with an environmental focus, help illustrate this. First is the apocalyptic forecast of journalist Robert Kaplan, who argues that environmental degradation will be "*the* national-security issue of the twenty-first century." [3] A litany of disasters—overpopulation, climate change, deforestation, soil erosion, water pollution and depletion, and so on—will loose floods of environmental refugees and thereby trigger violent political conflicts. An additional effect will be the spread of authoritarian regimes, especially in countries with well-established and influential military elites.

Kaplan believes these conflicts will lead to a "bifurcated" world dividing the Haves from the Have-nots. Both will be threatened by environmental decay and collapse, but only the Haves will be able to cope. Invoking historian Frances Fukuyama's notions of the end of history, Kaplan says the Haves will enter a "post-historical realm," insulated by their wealth and living, in effect, in gated communities and commuting to work in large (and polluting) air-conditioned limousines. The Have-nots, meanwhile, will suffer and struggle in bleak shantytowns and ghettos, gagging on the limousine's exhaust fumes and taunted by flickering television images of a prosperity that lies forever beyond their reach.

Who can help feeling a chill when reading this catalog of horrors possibly awaiting our children? By the same token, many will take comfort from the following vision of a technology-driven ride to an Edenic future, which we quote at length to give some sense of its exuberance:

> We are watching the beginnings of a global economic boom on a scale never experienced before. We have entered a period of sustained growth that could eventually double the world's economy every dozen years and bring increasing prosperity for—quite literally—billions of people on the planet. . . . In the developed countries of the West, new technology will lead to big productivity increases that will cause high economic growth. . . . And then the relentless process of globalization . . . will drive the growth through much of the rest of the world. . . .
>
> Five great waves of technology—personal computers, telecommunications, biotechnology, nanotechnology, and alternative energy— . . . could rapidly grow the economy without destroying the environment. . . . [Besides medicine,] the biotech revolution profoundly affects another economic sector—agriculture. The same deeper understanding of genetics leads to much more precise breeding of plants. By about 2007, most U.S. produce is being genetically engineered by these new direct techniques. The same process takes place with livestock. Superproductive animals and ultrahardy, high-yielding plants bring another veritable green revolution to countries sustaining large populations. . . .

The fifth wave of new technology—alternative energy—arrives right around the turn of the century with the introduction of the hybrid electric car. . . . By 2010, hydrogen is being processed in refinery-like plants and loaded into cars that can go thousands of miles . . . before refueling.

By the close of the 20th century, the more developed Western nations are forging ahead on a path of technology-led growth, and booming Asia is showing the unambiguous benefits of developing market economies and free trade. . . . By 1996, the [global growth] rate tops a robust 4 percent. By 2005, it hits an astounding 6 percent. Continued growth at this rate will double the size of the world economy in just 12 years, doubling it twice in just 25 years. . . . Almost every region of the planet . . . participates in the bonanza.

In 2020, humans arrive on Mars. . . . As the global viewing audience stares at the image of a distant Earth . . . the point is made as never before: We are one world. . . . By 2020, most people are acting on that belief. The population has stabilized. . . . Just as important, the world economy has evolved to a point roughly in balance with nature. . . . Rates of contamination have been greatly reduced, and the trajectory of these trends looks promising. The regeneration of the global environment is in sight.[4]

Obviously, both these scenarios are vulnerable to critique—the second, with its innocent faith in the saving power of technology, makes an especially tempting target—but that's not the point. Either one might come approximately true. Visions, like prophecies, can be self-fulfilling—like the bank whose rumored insolvency drives worried depositors to withdraw all their money, thus making the bank insolvent—or self-defeating, like the political candidate whose expected election win encourages supporters to stay home, thus leading to a loss at the polls.[5]

The purpose here is to show that visions of the future matter. It's not hard to imagine these sample visions either inspiring or alienating legions of impressionable young people. Moreover, sometimes we see things better by contrast, and visions can heighten the contrast between what is and what might be. They can also suggest starting points for effecting a transition from the one to the other.

Modern writers are not alone in exploiting the power of visions of the future. In exploring a way of approaching sustainability, this book puts a contemporary face on an ancient pastime: utopianism, or envisioning the world in a state of perfection. It is a practice that dates back at least to 2000 B.C. and the Sumerian epic *Gilgamesh*.[6] The unifying theme of utopian visions throughout history is how to provide for the common good in a world made unfair and unpredictable by both nature and human flaws. Utopian writings and experiments explore ways to arrange society to do a better job of promoting well-being. That is also the aim of the quest to achieve sustainability. The risk of global ecological tragedy adds a new and enveloping dimension to the problem, but the basic challenge is the same.

Utopian literature and history hold useful lessons for meeting that challenge. Perhaps the major lesson is that utopias invite trouble in paradise because they react to the inevitable failures of politics by trying to do away with politics. In sketching their visions, utopian writers generally thought that the ideal society could be defined as an end, when in fact perhaps all that can be defined is the means. Put another way, they focused on what the ideal society is, not on how to achieve it. This is a vitally important activity, and the modern tendency to launch immediately into problem-solving tends to slight it. Yet, the utopists—in essentially ignoring the How and simply wishing the What into existence—are also guilty of ignoring a crucial step. The discussion of utopianism that follows aims to show how that course does not serve the goal of sustainability.

Stories

"Utopia" is Greek for "no place." The word was coined for Sir Thomas More's book by the same name, published in 1516.[7] *Utopia* (discussed further below) became the most famous of writings on ideal societies, left its name to the field itself, and gave the English language a versatile synonym for "visionary," "idealist," and, sometimes, "fool." Most of the rich body of speculative writing on ideal societies comes after More's seminal work, as do most of the hundreds of attempts (apart from monasteries) to actually create ideal societies (see section titled "Experiments"). Perhaps not surprisingly, like the fictional Utopians and the monastic communities that often inspired them, nearly all these groups sought to isolate themselves from the world to one degree or another and to achieve some kind of communal life. Nearly all eventually failed, though a few lasted for decades.

Varieties of Utopian Experience

The human hunger for an ideal world has inspired at least five categories of such visions: Cockaygne, Arcadia, perfect moral commonwealth, millennium, and utopia proper.[8] Cockaygne is the English version of an ancient, persistent, and culturally widespread idea of unlimited plenty and leisure. In Cockaygne, every appetite of the human body, no matter how extreme, can be effortlessly sated. Food is abundant and delectable. The climate is calm and pleasant. The women are easy, the men forever young. Conflict does not exist, simply because there is no scarcity, thus no unmet needs, thus no need to fight over anything:

> In Cockaygne we drink and eat
> Freely without care and sweat,
> The food is choice and clear the wine,
> At fourses and at supper time,
> I say again, and I dare swear,
> No land is like it anywhere,
> Under heaven no land like this
> Of such joy and endless bliss.[9]

In this workingman's fantasy, noble lords and ladies must stand in pig muck for years to gain entry. In a sense, the promise of Cockaygne is much like the unspoken promise of modern technological capitalism: material plenty will eliminate scarcity and relieve us of the need to work out how to divide the limited goods and resources. In its time, however, everyone recognized the "Song of Cockaygne" as satire.

Arcadia is Cockaygne without the gross excesses, neither shortages of goods nor "longages of desire," in Garrett Hardin's words. Arcadia also displays greater harmony and balance between humans and nature. Nature is benign and generous, while people's needs are moderate rather than gluttonous. People live in relative ease and do not mind the light work they must do. Human society is so temperate and harmonious that the trappings of civilization are irrelevant. In his Arcadian essay "Of Cannibals," the sixteenth-century French writer Montaigne wrote of a nation

> in which there is no sort of traffic, no knowledge of letters, no science of numbers, no name for a magistrate or political superiority, no custom of servitude, no riches or poverty, no contracts, no successions, no partitions, no occupations but leisure ones, no care for any but common kinship, no clothes, no agriculture, no metal, no use of wine or corn.[10]

Arcadia is marked above all by the state of contented rest its inhabitants enjoy. One Arcadian fantasy, Gabriel de Foigny's *On Life and Death Among the Australians,* described the people there as preferring death to life, in the belief that there could be no more ideal state of rest than the grave.[11]

Very old conceptions of the third type of ideal society, the perfect moral commonwealth,[12] can be found among some of the early books of the Bible. The perfect moral commonwealth assumes that moral meaning for humans lies in their efforts to satisfy God's expectations; that is, to behave like saints. The so-called Hebrew literary prophets (Amos, Hosea, Isaiah, Jeremiah, and others) called this the messianic state.[13]

The idea of the perfect moral commonwealth was especially popular in early modern European thinking. In sixteenth-century England, it was widely believed that if every person, regardless of class or station in life, were morally perfected, perfect harmony—and thus order, stability, and happiness—would inevitably follow. Note that advocates of the perfect moral commonwealth, who were mainly members of the ruling elites writing for their peers, were not arguing against classes and stations. A class-based society was seen as the natural order of things.

In contrast to Cockaygne and Arcadia, the perfect moral commonwealth requires institutional arrangements among humans, but they work perfectly because the people who run them are morally perfect, or near enough. Thus, the Elizabethan writer Thomas Lupton could write of a place called Mauqsun (*nusquam* spelled backwards; see endnote 7) characterized by marvelous manners, honesty, courtesy, faithful friendships, true dealing, obedient wives and faithful husbands, modest maidens, diligent servants, and a general climate of mercy, charity, and Christianity. Ironically, there are lawyers in the perfect moral commonwealth—but only "loving" ones.[14] Apparently, no one wondered why lawyers would be necessary in a world of morally flawless humans.

Millennium, the fourth nonutopian category of ideal society, is the only one that pays any attention to the means of its achievement. In fact, that is its main focus. For many millennialists, only the Second Coming of Christ will launch the perfect society. The form and features of that society will be mostly up to Him and thus are not for humans to decide.

Millennialism, or millenarianism, was an immensely influential idea in seventeenth-century England. Some millennialists believed the Second Coming was imminent and should be approached with an attitude of expectant waiting. Others, more radical, argued that the arrival of the millennium could be accelerated by religious and political activism. This form of millennialism was feared and opposed by the Catholic Church from the days of St. Augustine in the fifth century, because of the challenge it posed to the Church's authority. Reformers like the Taborites (followers of radical preacher John Huss) in Bohemia and Savonarola in Florence had attempted to create communities governed by the "direct rule of God" in the 1400s, but these and other religious rebellions were violently suppressed with the Church's backing. Thirteen thousand Taborites were massacred in 1434,[15] and Savonarola was hanged, stoned, and burned with two companions in 1498.[16]

In general, the character of millennialist society was only sketchily described by those writing about it. Christ would reign and everyone would be holy. All Jews would be converted to Christianity. Laws might be needed . . . or they might not. There would be peace, health, and material plenty. The radicals often added lists of social changes they thought clearly deserved God's favor: making the law simpler and more accessible to laypeople, abolishing the national church and its tithes, land reforms.[17]

From these examples and brief descriptions, it is obvious that Cockaygne, Arcadia, perfect moral commonwealth, and millennium are all, to one degree or another, facile fantasies of ideal societies. Cockaygne simply assumes away any need for human morality or self-discipline, since nature meets any and all needs. Arcadia postulates a certain human temperance that is nicely in balance with nature's generosity but doesn't say how that temperance is to be inculcated and sustained. The perfect moral commonwealth argues that society can be made perfect by perfecting its members but is silent about where this tireless virtue would come from. And the millennialist vision essentially waits for God to solve the problem any way He chooses.

Genuine Utopia

Most writings about utopia proper offer more realistic solutions to the collective problem, to the extent that they generally begin with human beings as they are and then describe in detail the kinds of social arrangements and institutions necessary to achieve some ideal state. Utopias thus embody the smallest element of wishful thinking of any of the varieties of ideal society.

The purpose of the utopian state varies from one model to another. Some consider human happiness but all seek order and security, usually above everything else. The emphasis on order tends to lead to extensive regimentation, up to and including militarism. That tendency honors an old tradition, which can be illustrated by means of a closer look at three of the best-known utopias, Plato's *Republic*, Sir Thomas More's *Utopia*, and that modern homage to psychological engineering, B. F. Skinner's *Walden Two*.

Republic

It's been said that true utopian writings began with the publication of *Republic*, in about 394 B.C.[18] *Republic* was heavily influenced by the model of the Greek city-state Sparta, one of the most militaristic societies in history.

Sparta essentially existed for war. The city itself, which numbered seventy thousand at its peak, lay on the Laconia plain in Greece, surrounded by a natural fortress of mountains. From that stronghold, Sparta came to dominate its neighbors on the plain and then the rest of

Greece, despite often desperate measures of resistance. Historian Will Durant tells how

> the Messenian king, Aristodemus, consulted the oracle at Delphi for ways to defeat the Spartans; how Apollo bade him offer in sacrifice to the gods a virgin of his own royal race; how he put to death his own daughter, and lost the war.[19]

"Perhaps he had been mistaken about his daughter," Durant writes. "Two generations later, the brave Aristomenes led the Messenians in heroic revolt. For nine years their cities bore up under attack and siege; but in the end the Spartans had their way."[20]

The coming of the great leader Lycurgus marked a watershed in Spartan history. Before then, though warlike, the Spartans shared with their fellow Greeks a keen appreciation of culture and the arts. But sometime between 900 and 600 B.C., legend and tradition says, Lycurgus codified a body of laws and customs that rendered Sparta the rigid, militarized, and efficient conquering machine of story and song. Then, after exacting a promise from his fellow Spartans not to change anything until his return, he went away to Delphi and starved himself to death.[21]

The Spartans did their best to follow this difficult act. Spartan citizenship was based on landholding, but citizens were forbidden to acquire large holdings, since such wealth might lead to luxury and thus weakness. Every citizen from the ages of twenty through sixty was liable for military service. Travel and trade were discouraged for fear they would expose Spartans to corrupting influences. Men lived in dormitories and slept in the open, year-round, from childhood to the age of thirty. Celibacy was a crime, but marriage—instituted mainly for breeding—was highly arranged. If marriages could not be contrived for some adults, "several men might be pushed into a dark room with an equal number of girls, and be left to pick their life mates in the darkness; the Spartans thought that such choosing would not be blinder than love."[22]

Children were essentially the property of the state. Weak or sickly newborns were thrown off a cliff. Male seven-year-olds were taken from their families and formed into companies for their education and military training, which taught minimal literacy but stressed obedience and endurance of hardship and privation. Females' education concentrated on athleticism, to prepare them for the trials of childbirth.[23]

One contemporary observer remarked, only somewhat facetiously, that the Spartans were eager to die in battle because their lives were so miserable.[24] Still, Sparta's government and army were widely admired by many Greeks, including Plato. *Republic* describes Plato's notion of the ideal toward which rulers and states should struggle. The Greek philosophy of the time entertained several competing visions of the ideal state, but when *Republic* was written, Sparta had been the dominant city-state in Greece for almost two hundred years[25] and provided a compelling model.

As described in *Republic,* Plato's ideal state is limited to five thousand families, divided according to the "natural" order of things into the categories of the rulers and the ruled. Specifically, *Republic* splits the populace into four groups. On top are the lawgivers and policymakers (the Rulers) and, just below them, the managers and military commanders (the Auxiliaries). Together, these two subgroups are the Guardians. In the middle come the producers: farmers, artisans, and traders. At the bottom are the slaves.

Most of *Republic* is devoted to explaining the education and way of life of the Guardians, which begins with deep indoctrination designed to make them brave, unafraid of death, humorless, utterly self-controlled, and able to lie responsibly for the public good.[26] Once trained and on the job, Guardians may not marry or partake of the joys and comforts of family life. They may not own property, acquire wealth, or even secure an independent income. Their lives are austere and their needs, purposely kept minimal, are satisfied by the community. This enforced poverty is designed to promote unselfishness and utter devotion to duty.

Why would anyone want such a disagreeable job? In Plato's scheme, the incentives are mainly sex and respect. While alive, Guardians receive honor and recognition from their fellow citizens. (One reason for limiting the size of the state to five thousand families is that everyone can know the Guardians on sight.) They can freely sleep with women of their choosing, provided they are not carried away by lust. Above all, they have the ineffable satisfaction of performing the social role for which they are best suited and of knowing that the state is being served by those most qualified to lead. At death, they are given a suitable memorial.[27]

Modern politicians and autocrats might find these inducements unpersuasive. Nor is *Republic*'s society likely to appeal to anyone else, either. For example, Plato was suspicious of individualism and self-interest, especially as it was then being expressed in the rise of merchants and businessmen. Like many other agrarians, Plato despised merchants and saw them as driven only to "devote all their reasoning powers to calculating how to make money breed more money."[28] Plato's answer, since he believed that the scarcity of goods meant that one person's gain was another's loss, was to enforce egalitarianism so as to ensure harmony. Needs must therefore be limited, and the corrupting effect of excessive private property avoided.

Plato's republican state is completely static technologically and socially. The primary goal is a kind of freeze-frame unity and order achieved by regulations so elaborate and detailed as to seem Stalinist. For Plato, the business of the ideal society is security and the elimination of civic strife caused by competition for scarce goods. Though the title of his plan suggests at least some element of popular involvement in governance, in fact the overriding concern and means to these ends is top-down control. Plato and other utopists, as sociologist Charles Erasmus has pointed out, seem inevitably to discover that it's possible to maximize one value of a society in seeking perfection but not all values: "In fitting real societies to the rules of a utopia, the designer . . . must wittingly or unwittingly reduce them to police states."[29]

Utopia

The passage of time and the evolution of history did little to change that aspect of fictional utopias, if Thomas More's prototypical *Utopia* is a trustworthy guide. *Utopia* takes the form of a discussion among More, some friends, and a Portuguese traveler named Raphael Hythlodaeus (Greek for "dispenser of nonsense"),[30] who claims to have traveled around the world with Amerigo Vespucci and visited Utopia during the voyage. The book includes a blistering critique of social conditions in England

and, in the second part, Hythlodaeus' description of Utopia and the workings of its culture.[31]

The reader of that description is struck by the sameness and regimentation and by eyebrow-raising assumptions. Utopia's fifty-four city-states are all "spacious and magnificent, identical in language, traditions, customs, and laws." They do not compete for territory, "for they consider themselves the tenants rather than the masters of what they hold." The cities are much alike; streets are 20 feet wide and faced by houses with gardens at the back. The houses have doors in front and back, which admit anyone, being "easily opened by hand and then closing of themselves." The result is that "nothing is private property anywhere." Everyone wears essentially the same kind of clothing, which has not changed in design for centuries. "In Utopia a man is content with a single cape. . . . There is no reason . . . why he should desire more, for if he had them he would not be better fortified against the cold nor appear better dressed in the least." As for work, everyone (including women) learns agriculture and one additional craft, such as cloth-making, masonry, carpentry, or metalwork. That craft, unless special permission has been granted, is the one followed by one's father.

Behind a facade of democratic governance—election of group-household leaders, called phylarchs, and of city governors by the phylarchs—the watchword is regimentation. "The uniformity of life in Utopia has often been commented on," writes historian J. C. Davis. "Every aspect of the Utopian's daily life is subject to some form of regulation."[32] The cities are as identical as topography allows, the houses are all the same, everyone wears the same clothes, no one chooses his or her hours but works the same schedule as everyone else. The choice of work is not free, nor are choices of leisure. Discussing politics outside the senate is a capital offense. Travel outside the city-state without permission is prohibited and even then no one travels alone or where he wishes. Social relations are equally controlled: for example, marriage is forbidden for women younger than eighteen and men younger than twenty-two. There are no wineshops, alehouses, or brothels; also no "lurking hole, no secret meeting place." There is no privacy. Even suicides must have permission (though the punishment for violating this rule is not specified). People have little choice but to be "good":

While it is true that Utopians regard as valid only those laws "ratified by the common consent of a people neither oppressed by tyranny nor deceived by fraud," they have accepted a discipline which is totalitarian in its scope and denial of human individuality. If we mean by moral behavior a free choosing of the good rather than the bad when both alternatives are available, the Utopian's area of choice is so limited that he is almost incapable of moral behavior. In Utopia, the bad alternative is, so far as is possible, unavailable.[33]

More's argument for these constraints was that people respond rationally to the circumstances in which they live. Outside Utopia, he says, the circumstances (especially the emphasis on private property) conspire to force everyone to look first after their own personal interests at the cost of the public interest.[34] In Utopia, circumstances are arranged so that the public interest is uppermost in people's minds, and public well-being promotes individual well-being. "Assuredly in both cases they act reasonably," More argued—so the key was not to expect or assume saintliness in ordinary men and women, but to create a social incentive structure in which their rational actions would serve the public good. More thus anticipated the principles of behavioral engineering by several centuries.

Walden Two

The most prominent contemporary example of a fictionalized ideal community is B. F. Skinner's novel *Walden Two*, published in 1948. While *Utopia* gropes toward behavioral engineering, *Walden Two* is a full-blown behaviorist manifesto, written by the preeminent behaviorist of his time, the man who coined the phrase "the scientific study of behavior." *Utopia* implies that humans are creatures of reason and inner direction even though their conduct can be channeled predictably by arranging the circumstances in which they live. But *Walden Two* openly touts the study and manipulation of such arrangements as the key to paradise—and if that offends those who believe in the free will of human beings, so be it.[35] As Frazier, the founding genius, says, "I deny that freedom exists at all. I must deny it—or my program would be absurd. . . . Perhaps we can never prove that man isn't free; it's an assumption. But the increasing success of a science of behavior makes it more and more plausible."[36]

Walden (as it's called by the members) is a planned, communistic settlement of about a thousand people. Both production and consumption are collective. This arrangement extends to child-rearing, which is begun in communal nurseries and designed to weaken the parent-child bond. Walden replaces the family not only as an economic unit but also as a social and psychological unit. "Home is not the place to raise children," says Frazier, because parents, especially mothers, are too busy or distracted and because they really don't know what to do—and could not, without years of specialized training.[37]

Walden is run by specialized Managers, one for each of dozens of functional areas—health, the arts, dentistry, play, supply, advanced education, and so on. Managers are selected on merit, though it is not clear who does the selecting. They are not elected:

> "You work up to be a Manager" [said Frazier] "through intermediate positions which carry a good deal of responsibility and provide the necessary apprenticeship."
>
> "Then the members have no voice whatsoever," said Castle in a carefully controlled voice, as if he were filing the point away for future use.
>
> "Nor do they wish to have," said Frazier flatly.[38]

Skinner has left the top of the hierarchy for the Planners. They have a special role not given to Plato's Guardians or More's phylarchs. Walden is designed as a vehicle for scientific experimentation with the arrangements by which its people live together, and the work of designing the experiments is left to the board of six Planners. They are not elected, either; replacements are selected by the board from names supplied by the Managers.

Everyone at Walden is "happy." However, throughout *Walden Two* are various details of life at Walden that may or may not appeal to contemporary sensibilities. Some features—sheep keep the lawns cropped because lawnmowers are "the stupidest machines ever invented"—may be difficult to argue with, provided you prefer sheep droppings to noise, smoke, and fumes. On the other hand, daily life and interaction are governed by the

Walden Code, "certain rules of conduct . . . which are changed from time to time as experience suggests. . . . Each member agrees to abide by the Code when he accepts membership. . . . The Code acts as a memory aid until good behavior becomes habitual."[39] Examples of the Code include prohibitions against talking to outsiders about the affairs of the community and against gossiping about members' personal relations.

The governing arrangements, described briefly above, help Walden to be what Charles Erasmus calls a chessboard society:

> Both chess and utopia are stylized abstractions that provide a limited but graphic perfection. Utopian laws, like chess rules, however, are inflexible. A real state could no more be run according to the logic of utopia than a real battle could be won by following the logic of chess. The purpose of both games is to achieve a peculiar kind of freedom, like that of the laboratory or the theater, by withdrawing at least partially from reality.[40]

Walden is, in fact, a laboratory. In Skinner's vision, behavior is completely deterministic, and it is the task of science, as wielded by the Planners, to discover the laws of behavior and employ them to forge happy human beings. His view of inner-directed human potential is rather gloomy:

> "How could you give them freedom?" [asked Frazier.]
>
> "By refusing to control them!" [said Castle.]
>
> "But you would only be leaving the control in other hands."
>
> "Whose?"
>
> "The charlatan, the demagogue, the salesman, the ward heeler, the bully, the cheat, the educator, the priest—all who are now in possession of the techniques of behavioral engineering."[41]

Better leave it to the psychologists—that is, the experts, Skinner says. We are professionals; do not try this at home. "The people are in no position to evaluate the experts [says Frazier]. And elected experts are never able to act as they think best. They can't experiment. The amateur doesn't appreciate the need for experimentation."[42] And if the subjects of the experimentation disagree with Walden's policies or the Code? "Anyone may examine the evidence upon which a rule was introduced into the Code. He may argue against its inclusion and may present his own evidence. If the Managers refuse to change the rule, he may appeal to the Planners. But in no case must he argue about the Code with the members at large. There's a rule against that."[43]

The book never describes just how Walden and its thousand members are brought together, thus raising the bootstrap question of how a community with a radically different social ecology launches itself. By slow evolution, presumably, but the forces at work are not described. Did Frazier proselytize and recruit agreeable candidates? Did some flock to Walden hoping to fulfill their personal dreams of paradise, only to abandon it in disillusionment (a likely scenario, given the histories of many real intentional communities)? Were there purges? How did the community survive its birthing pains? We don't know. When we are introduced to Walden, it is a *fait accompli*, a completely formed and functional community with established traditions and culture. Skinner does not explain how to establish the foundation for this "good" life, except to assert that politics is certainly not the way:

> An important theme in *Walden Two* is that political action is to be avoided. . . . The great cultural revolutions have not started with politics. . . . What is needed is not a new political leader or a new kind of government but further knowledge about human behavior and new ways of applying that knowledge to the design of cultural practices.[44]

Experiments

Republic, Utopia, and *Walden Two* hardly exhaust all the designs and possibilities expressed in the utopian literature, but they are representative in that they are three of the most prominent and fully developed examples of the type. The outstanding lesson they present is their implausibility as serious options for a workable human society. They are implausible mainly because of their wishful assumptions about people's perpetual contentment under regimes of strict control.

Another way to put that assumption is that politics is made dispensable. In the chessboard societies of these and other utopias, there is no politics. Politics is seen either as unnecessary after material needs have been satisfied or as an impediment to the realization of the ideal conditions of utopia. The struggles, debates, negotiations, and maneuverings that constitute politics are therefore blithely assumed away in the construction of the ground conditions upon which these ideal societies are erected. *Walden Two* explicitly rejects politics, but all three works obviate them by presuming omnicompetence in the controlling classes and omnicontentment in everyone else.[45]

In addition, these three works describe societies that are essentially static. *Walden Two* is technologically progressive and assumes continued experimenting toward social perfection, but presumably some evolutionary pathways—such as eliminating the Planners and their role—are blocked by definition. The social structure is frozen. No one can know, of course, whether perfection would be self-sustaining, but does it matter? Observers from Heraclitus to Oscar Wilde have noted that change is the only constant. Not only is the world we know not static, one great truth of modern, postreductionistic science is that such highly complex systems as human societies generate *emergent* properties—that is, ones that cannot be foreseen no matter how complete the knowledge of preexisting conditions (see the next section). Much change is thus intrinsically unpredictable as well.

This is not to accuse Plato, More, and Skinner, all brilliant and worldly men, of naïveté. Plato and More's scenarios, at least, seem not to have been intended as realistic possibilities. Plato wrote *Republic* as an ideal toward which leaders should strive, and some scholars argue that *Utopia* is largely a vehicle for critiquing sixteenth-century English society. Skinner may be the exception; his faith in the power of the behavioral methods he championed reflected the scientific outlook of his time and seems absolute. Both faith and methods, however, were based on a belief in reductionism and have not been vindicated by social developments or more recent science. Behaviorism has declined in prominence in the field of psychology over the past fifty years.

In any case, these are fictions. What of the many attempts to create ideal communities in reality? Do they have anything different to say about the possibilities for transcending politics while defining and providing for the common good?

The short answer is no. The fictional utopias suggest that highly specified visions of a social structure require high levels of control. All three of these authors surely knew that the freewheeling turmoil of politics in a culture where power is widely dispersed was incompatible with their visions. The same theme shows up repeatedly in historical attempts to build ideal societies according to the precepts of a strong vision. Where those attempts succeeded, even for some time, there generally was or is a high degree of concentrated control or prescription, achieved by smallness and selection of members for compatibility and commitment.

For example, consider the Hutterites,[46] communist and pacifist farmers whose way of life has survived since its beginning in Moravia in 1528. As Anabaptists (Christians who believed that only adults should be baptized) in Catholic Europe, the Hutterites early and repeatedly felt the lash of religious persecution. By the eighteenth century, they had nearly been eliminated by war or reconversion to Catholicism. A handful eventually emigrated to Russia and then, in the 1870s, to the United States, where good farmers were in high demand to help settle the West. The American Hutterites prospered until World War I, when their pacifism and German background aroused jingoistic suspicion and hostility. After the deaths by maltreatment in prison of two Hutterite boys, most Hutterites left for Canada. Today, more than three hundred Hutterite colonies exist in North America, with perhaps 30,000 members.

That is a lot of people by the standards of intentional or alternative community movements. But the arithmetic reveals that the average Hutterite colony (called a Bruderhof, or "brother farm") has only about 100 members. This size preserves a familial atmosphere and helps prevent serious factionalism. When a community grows to 150 or so, it is divided, and half the members go off to start a new colony. Only senior males may vote on issues involving the colony's governance, so in a new colony of 50 members, only 6 may be allowed to vote, and in a mature colony of 150 members, only 18. Control is thus deliberately kept in the hands of a few.

Moreover, for all the virtues of this way of life—and Hutterites are intensely loyal to it—it is a highly prescribed one. Though Hutterites are progressive and technologically sophisticated farmers, they live only by farming; there are no other options. Leaving the community is difficult, for it involves a renunciation of one's "family" and the rejection of a tightly knit society that deliberately estranges its members from the mainstream. "To leave the colony structure is to belong to the devil and the outside world," said one runaway.[47] The role of women is traditional, and married women bear an average of eleven children each. New members are rarely accepted from outside and the group owes its continued growth to the successful indoctrination of children.

The 125 years of Hutterite commune survival in North America is unusual among alternative communities. The vast majority collapsed within weeks or months, though several lasted many decades. The Shaker communal movement began about 1800; as of 1995, one community of seven remained.[48] (Celibacy, a central feature of Shaker beliefs, takes its toll.) The Rappites were communists from 1805 until their dissolution in 1905, the Zoarites from 1819 until their demise in 1898. The famous Amana Society's True Inspirationists were communists for about ninety years, until the group converted itself to a shareholder corporation in 1932.

Like most intentional communities, these were small. Populations fluctuated over time, but they typically numbered from a few dozen to a few hundred members. Some Shaker communes may have briefly exceeded six hundred; none of the seven Amana Society villages contained more than five hundred.[49] Most of the communities were founded and headed by charismatic leaders with compelling visions, able to reinforce members' already-strong commitment to the goals and ways of the group.

This small size is critical, because an important consequence of commune smallness is social predictability.[50] When a group is small enough that everyone knows everyone else, then the group's affairs can be governed by social relations and expectations, which are continually shaped and reinforced by face-to-face interaction. The Hutterites' strategy of fissioning their colonies when they reach about 150 members is designed to conserve conditions of social predictability.

When a community grows larger, too many faces become unfamiliar, and it becomes difficult to know what to expect of others. Disputes arise that cannot be settled informally without a breakdown of order. At that point, successful management of the affairs of the community requires that conditions of legal predictability be established. Social relations and expectations must be codified and spelled out in laws and regulations. If someone violates those expectations, sanctions must be imposed by a civil governing body (city hall or the state). Social and legal predictability can overlap, but only in very small societies can the former dominate. The successful utopian societies achieved social predictability largely by remaining small and exclusionary.

Our fictional and experimental utopias must be small, regimented, or both, because purity excludes. The purer the utopian vision, the more people are left out. Whatever form human society takes in the decades ahead, however, it is unlikely that settlement patterns will devolve into millions of small, isolated communes of a few hundred members each. The trend is in the opposite direction, with people flocking to the cities, in part because most of them don't like very small communities: "It is the high visibility of the small community from which people have been escaping for hundreds of years. . . . It is a stifling world in which everyone monitors everyone else; it is the only real world in which Big Brother (the group) is forever watching each and every member."[51]

––––––––––

A lesson emerges from this look at fictional and real intentional communities: sustainability must rest on a foundation of pluralistic societies and systems of legal predictability. One can envision the purity of a utopia, but achieving it in the real world at a meaningful scale does not seem possible. We believe that progress toward sustainability will require sustainable communities to be globally predominant. Like a boat full of oarsmen drifting downstream toward a waterfall, a majority of the occupants need to pull upstream if they hope merely to stay in place, and real progress toward safety will require nearly everyone to row in unison away from the brink. Too many unsustainable, free-riding communities will drag the whole sustainability enterprise to a halt—and the default tendency of human society to become bigger, more populous, and more energy intensive will then threaten to carry all of us over the edge. Because of this need for a critical mass of participating people and communities, too much purity of vision—utopianism—is intolerable.

A second lesson is that we must assume and plan for continuous change, not only in society itself but in people's understanding of what is sustainable and their notions of the ideal society. Movement toward sustainability will have to be incremental and adaptive (open to feedback). Utopian purity of vision implies a static ideal, which inevitably would be left behind by the evolving real world and made irrelevant by events. What we should be seeking, therefore, is not utopia but *genotopia:* the place that is continually unfolding, being born and reborn.

These lessons point to a third: only in utopian fantasies is it possible to eliminate politics. The real choice is how to optimize our politics to make sustainability consciously approachable, and that requires a shift of political consciousness. Currently, economic growth is the canvas on which the world's cultures all paint their visions of the future. But that canvas is nearly full, and it's time to stretch a new one across the frame. The new canvas underlying those multiple pictures is sustainability, and the frame around the canvas is the ecological constraints on economic activity, which we discussed in chapter 2. We explore some of the key economic and social assumptions underlying our current politics and a way of making politics more responsive to the needs of sustainability in chapters 4 and 5.

NOTES

1. D. Meadows, "Envisioning a sustainable world," in R. Costanza, O. Segura, and J. Martinez-Alier, eds., *Getting Down to Earth: Practical Applications of Ecological Economics* (Washington, DC: Island Press, 1996), 117–126.

2. Ibid., 118.

3. R. Kaplan, "The coming anarchy," *The Atlantic Monthly,* February 1994, 58 ff.

4. P. Schwartz and P. Leyden, "The long boom: A history of the future, 1980–2020," *Wired,* online issue 5.07, July 1997 <http://www.wired.com/wired/5.07/longboom.html>.

5. R. Merton, "The self-fulfilling prophecy" (1948), in P. Sztompka, ed., *On Social Structure and Science* (Chicago: University of Chicago Press, 1996), 183–201.

6. I. Tod and M. Wheeler, *Utopia (An Illustrated History)* (New York: Harmony Books, 1978), 9.

7. More actually called the book *Nusquama;* "nusquam" is Latin for "nowhere."

8. J. C. Davis, *Utopia and the Ideal Society: A Study of English Utopian Writing, 1516–1700* (Cambridge: Cambridge University Press, 1981), 20 ff.

9. Cited in Tod and Wheeler, *Utopia*, 10, 11.

10. Cited in Davis, *Utopia and the Ideal Society*, 23.

11. Ibid. On at least one occasion, Arcadia was thought to be real, not just a visionary fantasy. In 1668, a compelling account appeared of a Dutch sea captain's discovery of an island "near to Madagascar." The Isle of Pines was named not for the native vegetation but for the colony of people found there, all descendants of one George Pine (then deceased) and four women shipwrecked with him. Pine described the pleasant and bounteous island as a paradise. Given the social circumstances, polygamy naturally ensued, and the castaways were fruitful. Life consisted mainly of sleep and love. Sixty years later, however, the population had grown to nearly 1,800, and paradise was lost. Resource shortages brought friction and competition, sin and civil war—and also the need for strict laws and a strong government to enforce them.

    The Isle of Pines caused a sensation. It appeared in multiple editions in English, Dutch, French, and Italian within a few months, and continued to be republished in Europe for one hundred years. But whatever the truth of its moral, the book eventually was proved a hoax.

12. Cited in Davis, *Utopia and the Ideal Society*, 26 ff.

13. J. O. Hertzler, *The History of Utopian Thought* (New York: Macmillan, 1923), 8.

14. Davis, *Utopia and the Ideal Society*, 29.

15. Tod and Wheeler, *Utopia*, 16.

16. Will Durant, *The Story of Civilization*, pt. 5, *The Renaissance* (New York: Simon and Schuster, 1953), 158–161.

17. Davis, *Utopia and the Ideal Society*, 35.

18. Tod and Wheeler, *Utopia*, 19.

19. Will Durant, *The Story of Civilization*, pt. 2, *The Life of Greece* (New York: Simon and Schuster, 1939), 73.

20. Ibid.

21. Ibid., 78.

22. Ibid., 84.

23. Tod and Wheeler, *Utopia*, 20, 21; Durant, *Story of Civilization*, 81–86.

24. Durant, *Story of Civilization.*

25. Ibid., 513, 86.

26. Tod and Wheeler, *Utopia,* 22.

27. C. J. Erasmus, *In Search of the Common Good: Utopian Experiments Past and Future* (New York: The Free Press, 1977), 207–217 passim.

28. Ibid., 200.

29. Ibid., 198.

30. This was one more clue to those of More's readers who knew Greek that the book was fiction, even though its detail convinced many it was a true account. Other clues included Amanote, a city name meaning "dream town," and the River Anider, which means "no water." See Tod and Wheeler, *Utopia,* 29.

31. Except as noted, the description and comments that follow are taken from More's *Utopia,* edited by Edward Surtz and published by Yale University Press (1964).

32. Davis, *Utopia and the Ideal Society,* 52.

33. Ibid., 54.

34. Two hundred fifty years later, Adam Smith would conclude that the individual pursuit of personal interest would lead, as if by an "invisible hand," to the collective interest. More saw the England of his day, with its poor-laws, hangings for petty thievery, seizures and enclosures of common lands by aristocrats, and other abuses, as the evil consequence of the intemperate pursuit of private interests.

35. *Walden Two* inspired several real attempts at intentional communities structured along Skinnerian lines, the most famous of which is Twin Oaks in Louisa, Virginia. Twin Oaks still exists thirty years after its founding, but its members rejected behaviorism long ago, along with many specific features of Walden, such as the variable labor credit system. For a highly readable historical account and memoir by one of its founders, see Kat Kincaid, *Is It Utopia Yet?* (Louisa, VA: Twin Oaks Publishing, 1994).

36. B. F. Skinner, *Walden Two* (New York: Macmillan, 1948), 241–242.

37. Ibid., 132.

38. Ibid., 49.

39. Ibid., 150.

40. Erasmus, *In Search of the Common Good,* 198.

41. Skinner, *Walden Two*, 240.

42. Ibid., 251.

43. Ibid., 152.

44. Ibid., xvi.

45. The Planners of Walden think they have an ace in the hole in this regard—
the ability to psycho-engineer contentment:

> "As to emotions—we aren't free of them all, nor should we like to be. But
> the meaner and more annoying . . . are almost unknown here, like unhappi-
> ness itself. We don't need them any longer in our struggle for
> existence . . . and it's easier to dispense with them," [said Frazier].
>
> "If you've discovered how to do that, you are indeed a genius," said Cas-
> tle. "We all know emotions are useless and bad for our peace of mind and our
> blood pressure. . . . But how arrange things otherwise?"
>
> "We arrange them otherwise here," said Frazier. ". . . It's simply a matter
> of behavioral engineering." (*Walden Two*, 92–94)

46. Unless otherwise noted, most of the information that follows about intentional communities is drawn from Tod and Wheeler, *Utopia*, and Charles
Erasmus, *In Search of the Common Good*.

47. Interviewed in the television documentary "Born Hutterite," directed by
Bryan Smith, July 1996. See <http://www.nfb.ca/E/1/2/bhut.html>.

48. D. Russo, "The last of the Shakers," *New York Times*, Current Events Edition,
May 7, 1995, 518.

49. Erasmus, *In Search of the Common Good*, 130.

50. Ibid., 46.

51. Ibid., 342.

Chapter 4

Prelude to Politics

The natural, seasonal flooding of the River Nile, essentially unmodified by humans until the nineteenth century, has supported a culture of settled societies since at least 5000 B.C.[1] Perhaps any society that has lasted seven thousand years should be entitled to call itself sustainable.

The Egyptian Nile valley culture, however, is a special case. The inhabitants were relatively few in number, and (notwithstanding the pyramids) their technology was modest. The annual deposits of silt that renewed the fertility of the soil resulted in part from deforestation and erosion thousands of miles upstream. That is, much of the carrying capacity of the Nile valley was imported; the culture had an "ecological footprint" larger than the valley itself. Moreover, settlement presumably occurred there so early, and has succeeded for so long, because the valley's natural features and processes make it particularly hospitable to habitation. In an empty world, the best places are settled first. If there were fewer of us and we lived differently, we, too, could settle only the best places, borrow a bit of carrying capacity from the rest, and so build a comfortable and sustainable life without doing much harm.

But there are 6 billion of us heading toward 9 or 10 billion over the next fifty years. We must occupy many places besides just the best. We live in a full or nearly full world, and the rules of the game have changed. In modern terms, a sustainable society must mean one that considers impacts at a global scale. It must mean one that has figured out not only how to tell when it is in balance with the supportive capacities of the whole Earth but also how to successfully apply that feedback so it can adjust itself to those capacities without the catastrophic intervention of nature. It doesn't count if the natural workings of famine, disease, and war slash population and consumption to sustainable levels. By that yardstick, sustainability is easy; continuing on the current track will achieve it sooner or later. For the goal of sustainability to be a meaningful one, we have to choose it consciously and then work toward it.

What sort of arrangement, though, is likely to be sustainable? As we have said before, this is a political issue that must be worked out within the limits described in chapter 2. In thinking about what a sustainable society might look like, two socioeconomic models deserve examination: capitalism and hunting-gathering. In many ways, they are polar opposites and represent extremes of worldview, organizational scale, wealth, purpose, spiritual tone, and quality of human relations. Capitalism dominates modern society in its roles as the soul of the conventional economic outlook and the soulmate of liberal democracy. It is possible to have autocracies with some measure of capitalism, as Singapore and even China show, and it is possible to have democracy without capitalism, as seen in the sampling of democratic societies in chapter 6. Nevertheless, in the modern world, capitalism and liberal democracy are so closely interwoven into one system that it can be difficult to distinguish their features. Every time a Wall Street investment banker moves over to run the U.S. Treasury, or the CEO of a Fortune 500 company becomes Secretary of State or Commerce—or they move back after a few years' "public service"—one struggles to see the line of separation. The machinery of liberal democracy seems increasingly to be the handmaiden of capitalism.

Ever since the collapse of the Soviet Union and the eastern bloc, capitalism has been treated as the inevitably ascendant socioeconomic form. Yet a great many ills can be laid at its door. Hunter-gatherer societies, on the other hand, have been almost completely expunged from the face of the earth, in large part by the expansionist forces of capitalism. This gives hunter-gatherers the cachet of the underdog and, as a historian once said, the supposed "superior virtue of the oppressed." Closer examination of the hunter-gatherer way of life reveals a more complex picture and some invaluable lessons. The one thing that capitalism and hunting-gathering have in common is that neither is sustainable.

Capitalism and Its Discontents

Capitalism is often briefly defined as "the private ownership of the means of production," but that is like saying a bordello is a hospitable place with many beds. Crucial information is missing. The dictionary definition of *capitalism* is more revealing: "The economic system in which all or most of the means of production and distribution, as land, factories, railroads, etc., are privately owned and operated for profit, originally under fully competitive conditions: it has been generally characterized by a tendency toward concentration of wealth, and, in its later phase, by the growth of great corporations, increased governmental control, etc."[2]

The key phrases in this definition are "originally under fully competitive conditions" and "concentration of wealth." It is debatable whether any capitalist economy has ever operated under conditions of full competition anywhere while simultaneously fulfilling the other conditions of ideal market systems; that is, perfect information among all buyers and sellers, mobility of capital and labor, absence of externalities (costs not captured in the prices of goods and services), freedom from distorting forces, participants who have no motivations other than maximizing personal interests, and so on. One of the myths of the market system is that capitalists prefer competition. To the contrary, though competition is celebrated by capitalism's apologists and benefits people (at least in their role as consumers) by keeping prices low, businessmen dislike it. They tolerate it only if they must, preferring to drive out, or buy out, competitors in order to control markets and maximize profits.

Nothing reflects the tendency toward concentration of wealth better than the large and increasing size and power of the great multi- and transnational corporations. This alone is troubling, since concentrations of wealth mean wide and unjust inequalities in wealth. But in addition, where wealth concentrates, so does power. One of the most chilling scenes in modern movies is in Norman Jewison's 1975 film *Rollerball*. It posits a future world completely controlled by a handful of monopoly corporations that distract the masses with the brutal ritual of rollerball. In this scene, the speaker, played by the forbidding John Houseman, heads the global energy corporation. He is hinting to rollerball player Jonathan E

(James Caan), a cult figure whose popularity has become a threat to the established order, why he must "voluntarily" retire:

> "Now we have the 'majors' and their executives: transport, food, communication, housing, luxury, energy. A few of us making decisions on a global basis, for the common good. . . . Corporate society takes care of everything. All it asks of anyone—all it's ever asked of anyone, ever—is *not to interfere with management decisions.*"

Are we headed for a *Rollerball* future? A capitalist dystopia, with or without democratic trappings, is not unimaginable, but neither is it inevitable. Capitalism's excesses and the current tendency to shrink before the task of fairly tempering them are not the final, fatal expression of human nature. Capitalism is an invented thing, not a "natural" thing.

Capitalism as an economic system has its very early roots in the gradual development of the commercial networks of northern Europe in the thirteenth and fourteenth centuries, when "trade, accumulation, and the reinvestment of profits into expanding and competitive business enterprises now became ends in themselves, not simply means to achieve personal wealth, land holdings, and aristocratic status."[3] But capitalism and the economic theory behind it rest on a set of assumptions about human nature that anthropologist Marshall Sahlins traces back to the story of the Fall, which essentially pegs human beings as creatures driven relentlessly by permanent and insatiable wants. Adam's curiosity, his ache for knowledge, condemned humanity to the perpetual seeking of satisfaction: "Both the cause and the crime consisted in the nature of man as an imperfect creature of lack and need. So did the punishment."[4] Banished from the Garden, where all needs were met effortlessly in the fashion of the Cockaygne paradise described earlier, humans were left scrambling for all eternity to satisfy their appetites.

According to Sahlins, this condition has undergone a profound reevaluation over time. The early Christian church saw it as humanity's permanent shackle, a form of bondage from which people could only be liberated by death in Christ and eternal salvation. Eventually, however, "this self-love changed its moral sign. The original evil and source of vast sadness in Augustine, the needs of the body became simply 'natural' in

Hobbes or at least a 'necessary evil' in Baron d'Holbach,[5] to end in Adam Smith or Milton Friedman[6] as the supreme source of social virtue."[7] Adam Smith, the eighteenth-century author of *Wealth of Nations* and one of the founders of classical economics, immortalized the phenomenon as the "invisible hand" of the market, by which the individual pursuit of self-interest leads to the greatest social good. Many scholars argued, in fact, that society itself, rather than having an organic source in the nature of humans as creatures of the group, was an instrumental result of the restless self-interest of isolated individuals: "Men congregate in groups and develop social relations either because it is to their respective advantage to do so or because they discover that other men can serve as means to their own ends."[8]

Thus was the intellectual groundwork for capitalism laid. It was reinforced by a complementary train of thought that developed during the Italian Renaissance. Sahlins describes Italian thinkers brooding about the source of value:

> [They] conceived capitalism as a total order of the universe well before it became a systematic economy. In 1440, for example, Nicholas of Cusa argued that human will and judgment were God's means of constituting the values of created things. Human preferences are the Deity's way of organizing the world as a *system of values*. . . . [Nicholas] thus prefigures the self-regulating market in the form of a cosmological process.[9]

Finally, the last piece in this puzzle-picture of human nature was that, although people were perpetually *and inherently* needy and unsatisfied, they were capable of making themselves into whatever they wished in the service of those needs and were not confined to a particular niche in nature. Other creatures, in contrast, were driven solely by the iron laws of instinct and hemmed in by the roles assigned them in nature's drama. This suggested the legitimacy of the view of "man as endowed with limitless possibilities of self-realization through the appropriation of nature's diversity"[10]—which might be called the anthem of the modern consumer.

These deep-rooted ideas about human nature are still reflected in many common attitudes people hold about themselves and others, including other races and the opposite sex; for example, blacks (or Chinese, or southerners, or the poor) are inferior, men stray because they can't help themselves, women's biology makes them unreliable as leaders, and so on. This biological determinism—the idea that nature and biology determine culture—is a particularly long-running and potent theme in Western thought:

> In people's existential awareness, cultural forms of every description are produced and reproduced as the objects or projects of their corporeal feelings. The system of the society is perceived as the ends of the individual. Not only kinship or college education but also Beethoven concerts or night baseball games, the taste of one Coke or another, McDonald's, nouvelle cuisine, suburban homes and *Picket Fences*, multimillionaire left-handed starting pitchers and the number of children per family, all these and everything else produced by history and the collectivity appear in life as the preferential values of subjective economizing. Their distribution in and as society seems a function of what people want.[11]

Capitalism then, if not without its excesses and transgressions, at least would *appear* to be profoundly in tune with the fundamental principles of our biology and mythology. We must admit, too, that capitalism excels at many good and useful things. By relying more heavily on markets than any other system, for example, capitalism generally solves the problem of allocation—how to distribute resources among various possible uses—more efficiently than other economic systems. The capitalist system creates an unequaled arena for innovation. In tandem with advancing technology and (extravagant) use of energy, capitalism has allowed a certain fortunate fraction of humanity to achieve undreamed-of material standards of living. Let it also be said that our quarrel is with capitalism as an overarching ideology of growth, expansion, and fetishistic moneymaking for its own sake and at whatever social and environmental cost, not necessarily with business per se, i.e., businesses whose purpose is to provide a useful good or service and thereby enable people to make a decent living.

The only major alternative to business and its profits as a source of employment is for everyone to work for the government.

But what about those excesses and transgressions? Are they simply the cost of our own nature?

Capitalism is so amply defended elsewhere that it is unnecessary for us to praise it further. Because they raise the question of whether we are getting our money's worth (so to speak) out of capitalism, its critiques are of greater interest. Those criticisms can be made at every level. For example, even people living in the developed world and enjoying the benefits of capitalism must confess how ugly its personal face can be. The internal dynamics of capitalism drive businesses to call you at home during dinner to sell you things, create artificial dissatisfaction and thus demand for junk, turn television into a sea of pandering trash and commercials, blight scenic areas with billboards, let the buyer beware, dare you to read the fine print, offer you Ponzi swindles, put your name and personal information up for sale, woo communities for the tax breaks and then abandon them when cheaper labor beckons elsewhere, and pretend frequent-flyer miles get you something for nothing, among many other deceptions, insults, and affronts. At a deeper and more systematic level, it is fatally blind to the critical issues of the scale of the global economy and the maldistribution of the world's wealth, denies ethical obligations to community welfare, shifts all possible costs to others (including the public), seeks to co-opt the political process by means of moneyed interest groups and otherwise erodes and corrupts the public sphere, and encourages the global homogenization of culture. Three of these problems are examined further below.

CAPITALISM NEEDS GROWTH. Capitalism is marked by what Robert Heilbroner calls a "consuming passion" for growth, of which the "bursting of corporate production beyond its national territory"[12] is only the most recent sign. This perception originated with Adam Smith, and there is no obvious disagreement about it among economists. Capitalism is, and always has been, inherently expansionist and self-enlarging, as firms seek to maximize profits in order to plow them back into productive assets to increase output, sell more, and thus further increase profits, ad infinitum. Capitalism, and the liberal democracies where it finds its most congenial homes, also rely on growth as ideology, since growth is the means to avoid the issue of capitalism's tendency to concentrate wealth, the resulting vast gap between the rich and the poor, and the need to devise fairer means of distribution. Capitalism's markets handle distribution with no thought to fairness by allotting goods and services according to the preferences of the world's consumers. Those preferences are expressed through consumers' buying power. Because poor people by definition have little money, in effect they have no preferences. They can command little obedience from the market because they have nothing with which to get its attention. The wealthy, on the other hand, can shout and scream at the market. (Right now in the United States, consumers are shouting for 3-ton, four-wheel-drive trucks, $5,000 gas grills, expensive wristwatches, premium wines, and other conspicuous consumption goods. U.S. sales of such luxury goods are up sharply in recent years.)[13] To appease those who stand to gain by redistribution, it is necessary to promise a continually enlarging economic pie rather than devising ways of fairly apportioning existing wealth among all players.

But we already know the trouble with perpetual growth: it's impossible. Economic production, which capitalism excels at, is the process of turning natural capital (ecosystem resources) into manufactured capital, while finding somewhere to put all the wastes that production entails. Since the global ecosystem is both the source of the inputs for production and the sink for the wastes, and since the ecosystem is finite, obviously at some point the conversion process must stop expanding. Long before that hard limit has been reached, though, ecosystem degradation and the associated human costs will have become punishingly severe.

CAPITALISM CORRODES THE CIVIC MORAL FOUNDATION IT NEEDS TO FUNCTION. The readiest example of this is the obnoxious salesman. You arrive at a vacation resort, anticipating a few days of carefree relaxation, and within seconds of stepping out of the cab are accosted by a salesman for time-share condominiums. He plants himself athwart the sidewalk and launches into his pitch. Despite demurrals, he keeps talking and refuses to let you pass. You edge around him; he follows you down the street. The exchange finally ends in epithets. By exploiting your civility to command your attention, the salesman leaves one more minuscule crack in the civic moral foundation capitalism depends on.

Advertising offers other examples of this effect. Heilbroner called advertising "perhaps the single most value-destroying activity of a business civilization," one whose chief effect was to "teach children that grownups told lies for money."[14] One way it does this is to corrupt language for the ends of the business culture. The overuse of the word "new" comes to mind immediately. Other examples include "pre-owned" cars, and "homes." Once upon a time, a house was not a home. Before the real-estate industry cranked up its well-oiled advertising machine, a "house" was a physical building designed to shelter a few people, while a "home" was a psychological construct built up out of the experience of living in the house, the associations to friends, neighbors, and community, and so on. Developers built houses and real-estate agents brokered them, but it was families who made houses into homes. Since the real-estate business expropriated the word "home" for its positive connotations it is the homes that are built and sold. They would have us believe, at some level, that what the inhabitants once created by investing their hearts and souls can now simply be bought. An ancient word with powerful emotional resonance has been turned to the service of a private, narrow interest. "How strong, deep, or sustaining," Heilbroner asked, "can be the values generated by a civilization that generates a ceaseless flow of half-truths and careful deceptions?"[15]

Capitalism depends on certain "conditions of production," including an educated labor force, natural resources and supportive ecosystems, and infrastructure, such as roads, airports, and water and sewer systems. Capitalism itself does not supply these conditions, it merely uses them and/or seeks to control them. In fact, it tends to use them up, leading to what Marxist critics have called the "second contradiction" of capitalism:[16] the cost-side crises triggered by neglect of the conditions of production in order to preserve or maximize profits.[17] As they try to reduce costs or shift them to others, firms have a tendency, for example, to dump wastes cheaply and expeditiously and thus degrade water and air quality, extract resources without making provision for their restoration, skimp on workers' health care and thereby compromise the health of the work force, and so on. Eventually, the firms find themselves without the inputs needed for production.

Besides labor, resources, infrastructure, and so on, the conditions of production include social orderliness, human fellow feeling and sense of community, and other "extra-market values." Without them, capitalism could not function:

> Norms of civility are a public good. Without them, the world would degenerate into a society of relentless mutual suspicion. The late Olof Palme used to warn of the dangers of a "society of sharp elbows." Instead of a relatively pleasurable world of commerce, we would have to be constantly on guard against being ripped off. Bankruptcies would proliferate. Banks would have to charge higher interest rates to compensate for the ubiquity of opportunism. Explicit warranties would have to be negotiated for every transaction. Doctors would resort to "defensive medicine" to a far greater degree than they already do. A handshake would become worthless. Society would be a paradise mainly for lawyers. This is the dystopia the marketizers commend.[18]

In a thousand blatant and subtle ways, capitalism chips away at the foundation of the extra-market values it rests on.

CAPITALISM CORRUPTS THE PUBLIC SPHERE. A related flaw has to do with capitalism's relationship with the public sphere, which is the psychic and political space and process within which people, acting as citizens, consider their common dilemmas and seek solutions. Political self-determination depends on the existence of a robust public sphere. The essence of that robustness is ongoing debate in which truth is determined by reason, not by status or power, and which involves continual critique-driven evolution of argument. A strong public sphere is the only alternative or counterweight to markets and state authority (money and power) as a means of ordering human social life.[19] Capitalism initially helped to create the middle-class form of the public sphere, but then helped to destroy it.

In the West, the fortunes of the public sphere have waxed and waned over the centuries. The ancient Greeks had a vibrant public sphere, while the Roman tendency toward empire and autocracy suppressed it. Feudalism was not congenial to a true public sphere either, but the late seventeenth and early eighteenth centuries saw its revival in Europe among the middle classes, where, in the words of sociologist Jürgen Habermas, "Private people came together as a public" and "soon claimed the public sphere regulated from above against the public authorities themselves, to engage them in a debate over the general rules governing relations in the basically privatized but publicly relevant sphere of commodity exchange and social labor."[20]

Habermas described this flowering of the public sphere in stupefying detail in his classic study *The Structural Transformation of the Public Sphere*. Briefly, in the seventeenth and eighteenth centuries the proliferation of coffeehouses as meeting and debating places (three thousand of them in London alone by about 1710) brought together a surprisingly wide variety of people, especially from the middle class, and encouraged critical discussion of literature and art and, later, economic and political issues. Coffeehouses, salons, reading societies, and other meeting institutions in Britain, France, and Germany helped "preserve a kind of social intercourse that . . . disregarded status" and helped define "the public" more inclusively, as well as wresting away from church and state authorities exclusive control of many subjects that were considered to be common concerns. The discussion was abetted by the invention of many critical journals and moral weeklies and of literary and political journalism in general. The development of capitalism intertwined with this process, as local economies expanded to become regional and then national in scope by means of widened trade. This created a need for news and information about prices and demand. Periodicals (many of them owned by banks) that started out to supply those needs soon began carrying other information that fueled public discussion about important issues.[21]

The public sphere in this era actually included only a small minority of the total population. From the 1750s, conflicts among moneyed interests—commercial and trading capital on the one hand and manufacturing and industrial capital on the other—helped widen the sphere somewhat, since the various parties in the conflicts saw advantage in seeking allies wherever possible. In practice, though, the public sphere was "the bourgeois reading public." To be a member required education and property, conditions that generally excluded females. In principle, though, under the assumptions of classical economics, everybody was assumed to have a more-or-less equal chance, with ability and luck, of attaining the required education and property and thus a legitimate place in the public sphere.

The public sphere's influence continued to deepen and became so important that by the 1800s the public "in the role of a permanent critical commentator . . . had definitively broken the exclusiveness of Parliament and evolved into the officially designated discussion partner of the [members of Parliament]. . . . 'They,' the subjects of public opinion, were no longer treated as people whom, like 'strangers,' one could exclude from the deliberations. Step by step the absolutism of Parliament had to retreat before their sovereignty."[22]

This period did not last, however. All this was taking place in the context of the rapid development of the market economy and the social turmoil and dislocation it produced. The fiction of equal access to the bourgeois classes for those whom the market economy disadvantaged could not be kept up. In the nineteenth century, more groups agitated successfully for expansion of the franchise, a process that went on for decades. But as the public sphere widened further, so did conflict. The very class divisions that the public sphere once had suspended, at least in principle, became themselves the subjects of debate. In its classical phase, the public sphere saw itself as debating to achieve some kind of consensus on what was in the common interest, not to achieve power. Later, however, public opinion became divided and coercive, as whichever segment held the upper hand sought to work its will. The public sphere became an arena within which different interest groups vied to assert rights and seek state protection of those rights. Disillusioned political thinkers like J. S. Mill and Alexis de Tocqueville came to think of public opinion as "the reign of the many and the mediocre" and concluded that "public opinion determined by the passions of the masses was in need of purification by means of the authoritative insights of materially independent citizens"; i.e., representative government by elites. "Educated and powerful citizens were supposed to form an elite public . . . whose critical debate determined public opinion."[23]

The basis for the public sphere thus began to erode. With the further development of capitalism, it also became obvious that the classical liberal idea of an economy of free competition and independent prices was a fantasy. In the real world, some economic actors almost always had better information than others, or were able to influence market prices by buying or selling large quantities of a good, for instance. Under these and other conditions of market imperfection, power and wealth were increasingly concentrating. As this tendency developed, it was the forces for democratization that saved capitalism from itself by interceding on behalf of workers. The mechanism was increased state intervention. In effect, though, this growth of power within the spheres of capital and the state tended to crush the old public sphere between them. Eventually, the deflation and collapse of the public sphere became pronounced: although it penetrated more *areas* of society, "It simultaneously lost its political *function,* namely: that of subjecting the affairs that it had made public to the control of a critical public."[24]

Parallel social developments aided this process. The basis of the old idealized public sphere was that it was composed of middle-class, autonomous individuals. Their autonomy came from their property ownership and consequent independence—Britain as "a nation of shopkeepers," for instance—within a market economy that operated privately; that is, free of interference from church, state, and aristocracy. But over time, business and state organizations grew larger and more impersonal and began to interpenetrate each other more and more, through regulation, for example. The public sphere also widened to include many wage earners, people without property who made their livings not by owning some small means of production but by working for someone else. The once-private world of work developed into something like the realm we experience today, separate from both home life and public life. The realm of what was "private" lost the economic part of its meaning and came to mean only the family. Consequently, "The public sphere was turned into a sham semblance of its former self. The key tendency was to replace the shared, critical activity of public discourse by a more passive consumption on the one hand and an apolitical sociability on the other."[25] Lost was any strong sense of a common public interest commonly defined and worked toward, replaced by a sense of individual and group interests and rights that were to be negotiated.

It is not too many more steps to the current situation: on the one hand, we have a society of mass consumerism, full of people who have abandoned public discourse to concentrate on swallowing the endless stream of products and entertainment generated by huge capitalist organizations. On the other hand, we have a professional political class, including journalists, heavily influenced by corporate lobbies and organized interest groups; polling, public relations, and professionalized argumentation on political talk shows ("Crossfire," "Meet the Press," etc.) as substitutes for true public debate; low and declining voter turnout; and widespread alienation from the political process. Especially at the national level, the public sphere in its classical sense has almost ceased to exist.

Critiquing capitalism has been a favorite pastime for hundreds of years, and the brief arguments recounted above are not new. But they bring us to an important question: capitalism may be troublesome, heedless, and self-destructive, but can't it be fixed? Or is there something inherent in human nature that leads inevitably to the behaviors of individuals (and thence organizations) that create the fundamental character of capitalism?

The individual seen through the lens of capitalism is called *Homo economicus,* economic human. *Homo economicus* always wants more goods, is always rational and consistent in his or her choices, always has perfect information about goods and prices, and always acts to maximize personal satisfaction. That satisfaction comes only from things he or she has personally consumed (or paid for, in the case of gifts given to others). *Homo economicus* does not vote and cares nothing about the beauty of a smog-free sunrise or the honest regard of family and peers. *Homo economicus* "takes no pleasure when a neighbor receives a gift from someone else, or when a philanthropist endows a park for underprivileged children. Similarly, *Homo economicus* is not envious of the neighbor's new car or pained by defeat in competition for an honor. *Homo economicus* knows neither benevolence nor malevolence in any of these instances, only indifference."[26] Human beings in this model are first, last, and always extreme individualists: economic molecules, in the conception of Leon Walras, one of the founders of modern economics.

Homo economicus is obviously one of those pesky simplifying assumptions economists make in order to construct a manageable theory. The trouble with this assumption is that it is seriously flawed, as well as fatalistic. Real people are far more complex (and interesting) than *Homo economicus*, a fact that a few economists are beginning to grapple with in their research. In the embryonic field of behavioral economics, researchers studying actual human beings in experimental and real-world situations have, for example, discovered the following:

- Contrary to conventional economic theory, although the most logical and profitable strategy for winning at the track is to bet on the top three horses, people often bet on long shots (a consistently unprofitable strategy) more than half the time.
- People are loss averse; they will pay far more to avoid losing something they value than they will pay to secure it in the first place, a tendency that makes no sense in conventional economics.
- Faced with the choice of a fifteen-minute break today or a thirty-minute break tomorrow, workers generally choose the short break today. If asked to choose between a short break 100 days from now and a longer break 101 days from now, though, most will choose the long break. At that remove, an extra day doesn't matter. Standard economics predicts the second outcome but not the first; reckoning without the vagaries of human patience, it says the thirty-minute break should always be the top choice.[27]

Even real human beings steeped in the ways of capitalist society, such as ourselves, differ in critical respects from the characteristics of *Homo economicus*. This alone seriously undermines the theoretical foundation upon which the capitalist edifice rests, for that foundation is made of assumptions about the intrinsic nature of human beings that turn out to be porous. But what if someone had scouted around and discovered groups of people whose behavior and social characteristics utterly rebuked the conventional model? Wouldn't that shatter the capitalist foundation irrevocably?

Behold the hunter-gatherers.

A-Hunting We Will Go

"There is a passion for hunting something," wrote Dickens, "deeply implanted in the human breast." For hundreds of thousands of years, if not since our hominid ancestors swung down out of the trees, human beings have hunted and gathered. In the modern era, it is a rapidly vanishing way of life, even in the least-developed parts of the world. In the industrialized world, our hunting is mainly for bargains and parking places, and our gathering takes place at the grocery store.

The classic formulation of life among primitive peoples comes from Thomas Hobbes, writing in *Leviathan* (1651): "No arts; no letters; no society; and which is worst of all, continual fear and danger of violent death; and the life of man, solitary, poor, nasty, brutish, and short." For decades, modern anthropological research supported the stereotype:

> "Mere subsistence economy," "limited leisure save in exceptional circumstances," "incessant quest for food," "meager and relatively unreliable" natural resources, "absence of an economic surplus," "maximum energy from a maximum number of people"—so runs the fair average anthropological opinion of hunting and gathering.[28]

Hobbes's famous summary is an arresting and memorable description of primitive life. Its only shortcoming is that it is all wrong. Research conducted since the late 1960s on archaeological evidence and on remaining hunting and gathering groups, such as the Hadza, Bushmen, Inuit, and Australian Aborigines, has led to a radical revision of the scientific picture of hunter-gatherer peoples. Far from living on the perpetual edge of starvation, the collective generations of them listlessly awaiting, so to speak, the salvation of settled agriculture, hunter-gatherer peoples that were not forced into the poorest habitats lived relatively comfortable lives without working very hard. In other ways, too, they offer a striking and instructive contrast to our own, "superior" societies based on agriculture and industrialism. Following are some of the typical features of that way of life, observed, reconstructed, or deduced from studies of living groups and the archaeological record.[29]

MOBILITY. If the berries and the wildebeest won't come to you, you must go to them. Most hunter-gatherer groups are nomadic. Small groups of perhaps one to several dozen people live in temporary camps, with several camps sharing a territory large enough to support all. The camps move frequently, and there is often considerable mobility between them; that is, people can move from one camp to another without economic penalty.

NONMATERIALISM. An interesting result of bipedal nomadism is that possessions are a burden. Few groups have draft animals, and without them the hunters must carry everything they own when it comes time to move camp. Consequently, they own very little. More to the point, they do not lament this condition; they do not see it as poverty. If they lack material goods, it is not because the demands of the hunt and the search for edible plants are so severe (on the contrary; see below), but because they are culturally inclined against such goods. Marshall Sahlins puts it like this:

> The hunter, one is tempted to say, is "uneconomic man." . . . He is the reverse of that standard caricature immortalized in any *General Principles of Economics*, page one. . . . It seems wrong to say that [his] wants are "restricted," desires "restrained," or even that the notion of wealth is "limited." Such phrasings imply in advance an Economic Man and a struggle of the hunter against his own worse nature, which is finally then subdued by a cultural vow of poverty. The words imply the renunciation of an acquisitiveness that in reality was never developed, a suppression of desires that were never broached. . . . It is not that hunters and gatherers have curbed their materialistic "impulses"; they simply never made an institution of them.[30]

RELATIVELY GOOD HEALTH. Though they do not benefit from the elaborate and costly health-care systems of industrialized nations, neither do hunter-gatherers suffer from the host of modernity-related diseases that plague moderns. Nutrition is generally rather good, with a wide variety of foodstuffs available for the taking. Caloric intake is more than adequate, and diseases of malnutrition are rare. Exercise is plentiful. Except in vegetation-poor arctic areas, gathering provides most of the group's food, since hunting is a difficult and relatively unrewarding business. Life expectancy and infant mortality rates among early hunting and gathering groups, while certainly not as favorable as those in modern developed nations, were apparently at least as good as those in early agricultural societies.[31]

ABUNDANT LEISURE. Securing these benefits does not require a hard work week. On average, those who work do so, intermittently, for a total of three to six hours a day—but not every day. (The total work "week" of the Dobe Bushman of southern Africa is about fifteen hours.) And not everyone works; about a third either are not expected to (the young and the old) or simply choose against it.

EGALITARIANISM. Although not all hunter-gatherer groups display equality of the sexes and a lack of social stratification, these are strong features in many of them, especially those living in what are called immediate-return systems.[32] In immediate-return systems, food is generally consumed within hours or days of its collection (and women, as the chief gatherers, enjoy considerable economic power). In contrast, delayed-return systems (including agriculture) entail food processing and storing and often much more elaborate tools, implements, permanent dwellings, and other assets. Ownership of these production assets is carefully defined and controlled; people interact in a matrix of rules and quasi-legal relationships. One category of assets is the men's female relatives, who can be given in marriage to other males. No such arrangements exist in immediate-return groups, where the norms include greater flexibility in social groupings, more choice in personal relationships, and an emphasis on sharing. As for hierarchies, the lack of important possessions precludes stratification based on wealth, and immediate-return groups either have no leaders, or the leaders have very limited powers. In some groups, a rule of thumb is that anyone desiring power is almost automatically prevented from acquiring it.

SOCIAL SECURITY. In most societies, those who do not work do not eat. That is the underlying sentiment, at least, though its strength varies from place to place, and most developed nations construct welfare safety nets. Capitalist economic systems closely link consumption with production. Indeed, they are predicated on the idea of pay commensurate with the value of one's output. Those availing themselves of the safety nets, rather than producing, are often disdained and sometime reviled. In many hunter-gatherer cultures, on the other hand, there is no connection between production and consumption. Some members literally do nothing productive at all, for years on end, yet are routinely given, ungrudgingly, their full share of the available food.

ECOLOGICAL SENSITIVITY. Since their welfare depends directly upon it, it is not surprising that hunter-gatherers typically display profound and detailed knowledge and understanding of the ecologies where they live. Hunter-gatherers are not always the ecological saints some writers have made them out to be; there is clear evidence of heedless environmental destruction among hunter-gatherers at various times and places, ranging all the way up to wholesale forest clearing and the extinction of entire species. Nevertheless, their modest needs, low numbers, and lack of "advanced" technologies have enabled many of them to persist as distinct cultures for tens of thousands of years without doing serious ecological damage.

———

The lesson of the hunter-gatherers is not that their way of life can save us. The world is too full now for all of humanity to live that way, even if we could unambivalently jettison the hopes and pleasures of modern life. Just as capitalism ultimately is not sustainable because its insatiable hunger for growth destroys the ecological body of Earth and because its values corrode the civic body politic it needs to function, neither is hunting and gathering ultimately sustainable, because of its own ecological interference and because it succeeded so well that human populations expanded to the point of requiring the wholesale conversion to agricultural modes of output[33]—the early basis for surpluses and thus property, rights over that property, haves and have-nots, and finally capitalism itself. In both cases, necessary feedback mechanisms and effective means of socially constraining excesses were lacking. We conclude that, in terms relevant to modern conditions, no society has ever succeeded in achieving a sustainable economic and social structure that will work globally into the indefinite future.

However, the hunter-gatherers do offer a compelling lesson; that is, that human beings unschooled in the ways of modern capitalist society are not naturally hierarchical, materialist, acquisitive, territorial, overly competitive, misogynistic, or power mad. The solitary, poor, nasty, brutish, and short lives attributed by Hobbes to primitive peoples are more likely to be found in the permanent underclass, the detrital victims of capitalist systems. The claim that *Homo economicus* embodies the essential truth about human beings is revealed not only as a delusion but as a snare as well. Since ancient times human beings have been shaped by their cultures, not struck from a mold that thereby creates culture in its image:

> Culture antedates anatomically modern man *(H. sapiens)* by something like two million years or more. Culture was not simply added on to an already completed human nature; it was decisively involved in the constitution of the species, as the salient selective condition. The human body is a cultural body, which also means that the mind is a cultural mind.[34]

To put it another way, humans are malleable and highly social beings. Biology helps shape us, but biology is not destiny. The lesson of the hunter-gatherers is that humans can be anything that makes sense. If it no longer makes sense—social, ecological, or political—for us to behave as the economic molecules that capitalism falsely claims is our instinctive fate, then we can be something else. What that vision of human potential should be, and how best to move toward its fulfillment, are the subjects of politics, to which we now turn.

Notes

1. C. Ponting, *A Green History of the World: The Environment and the Collapse of Great Civilizations* (New York: Penguin Books, 1991), 83, 84.

2. *Webster's Deluxe Unabridged Dictionary,* 2nd ed. (New York: Simon and Schuster, 1979), 269.

3. M. Bookchin, *From Urbanization to Cities: Towards a New Politics of Citizenship* (London: Cassell, 1992), 118.

4. M. Sahlins, "The sadness of sweetness: The native anthropology of Western cosmology," *Current Anthropology* 37(3) (June 1996): 397.

5. Baron d'Holbach was one of a group of eighteenth-century writers and thinkers that included Montesquieu, Rousseau, and Voltaire.

6. Adam Smith was the author of *Wealth of Nations* (1776), the founding manifesto of classical economics. Milton Friedman is perhaps the most famous member of the so-called Chicago School of conservative economic theory.

7. Sahlins, "The sadness of sweetness," 398.

8. Ibid.

9. Ibid., 399.

10. Ibid., 400.

11. Ibid., 401.

12. R. Heilbroner, *Business Civilization in Decline* (New York: W. W. Norton, 1976), 107.

13. R. Frank, "Our climb to sublime," *The Washington Post,* January 24, 1999, B1.

14. Heilbroner, *Business Civilization in Decline,* 113, 114.

15. Ibid., 114.

16. J. O'Connor, "Is sustainable capitalism possible?" in M. O'Connor, ed., *Is Capitalism Sustainable? Political Economy and the Politics of Ecology* (New York: Guilford Press, 1994), 160.

17. The "first contradiction" of capitalism is the demand crisis that results from firms' attempts to reduce wages, improve productivity, and lay off workers in order to compete more effectively. They thus achieve greater output at lower costs, but they also reduce the number of people who can buy their products. Higher output coupled with lower demand drives prices down and thus squeezes profits.

18. R. Kuttner, *Everything for Sale: The Virtues and Limits of Markets* (New York: Knopf, 1997), 66.

19. C. Calhoun, *Habermas and the Public Sphere* (Cambridge, MA: MIT Press, 1992), 6.

20. J. Habermas, *The Structural Transformation of the Public Sphere: An Inquiry into a Category of Bourgeois Society* (Cambridge, MA: MIT Press, 1989 [1962]), 27.

21. Calhoun, *Habermas and the Public Sphere*, 8.

22. Habermas, *Structural Transformation of the Public Sphere*, 66.

23. Ibid., 137.

24. Ibid., 140.

25. Calhoun, *Habermas and the Public Sphere*, 22–23.

26. H. Daly and J. Cobb, *For the Common Good: Redirecting the Economy toward Community, the Environment, and a Sustainable Future* (Boston: Beacon Press, 1989), 85–86.

27. B. Daviss, "Let's get emotional," *New Scientist*, September 19, 1998, 39–41.

28. M. Sahlins, "The original affluent society," reprinted in J. Gowdy, ed., *Limited Wants, Unlimited Means: A Reader on Hunter-Gatherer Economics and the Environment* (Washington, DC: Island Press, 1998), 6.

29. For elaborations of these points, see Ponting, *A Green History of the World*, 18 ff., 32–35, 225, and Gowdy, *Limited Wants, Unlimited Means*, especially chapters 1, 4, and 5.

30. Sahlins, "The original affluent society," 15.

31. Ancient peoples may have lived much longer than once believed. Recent studies suggest that the statistical methods used to estimate age from the condition of skeletal remains may underestimate age at the time of death by as much as thirty years. See R. Matthews, "Digging the dirt on age," *New Scientist*, March 13, 1999, p. 18.

32. See J. Woodburn, "Egalitarian societies," in Gowdy, *Limited Wants, Unlimited Means*, 87–110.

33. Ponting, *A Green History of the World*, 41, 42.

34. Sahlins, "The sadness of sweetness," 403.

Chapter 5
Engaging Politics

Does democracy promote sustainability? It is tempting to conclude that it does, if only because the environmental records of the world's democratic nations generally look better than those of autocratic nations. On the other hand, perhaps democracy is not what makes democratic systems seem more sustainable. Many democratic nations are also wealthy, and wealth (high per capita incomes) is thought to correlate both with strong popular environmentalism and with lower levels of pollution.[1] The theory is that people become able to act on their environmental concerns only when their disposable incomes exceed a certain threshold. The poor may care just as much about the natural world around them but are locked in a struggle for survival and are unable to act on those beliefs. So the environmentalism of the North may be merely an artifact of its riches. Because much of that wealth is the legacy of aggressive colonial expansion that in effect borrowed or stole wealth from the rest of the world, perhaps the environmentalism of North America, Europe, and Japan has been bought at the expense of the impoverished South. If so, the South cannot emulate the northern strategy for achieving sustainability.

The correlation between democracy and environmental progressivity is only suggestive, not conclusive. But there are also good theoretical reasons to believe that autocracy cannot match democracy's sustainability potential. First, through their ability to corner the available wealth and power, the elites that rule in autocratic countries can ignore or insulate themselves from all kinds of problems, including those that stem from environmental degradation. Second, problems requiring collective action are more effectively addressed when civic collaboration is possible, and the vertical networks characteristic of autocracies are much less effective in facilitating such collaboration than the horizontal networks of democracies.[2] Third, to hold on to their power and privileges, autocratic elites attempt to suppress or eliminate conflict, yet bounded political conflict is necessary to detect and correct errors of policy and perception.[3] Wealthy rulers whose power is uncontested do not have to listen to anybody except cronies and hangers-on (because they can simply jail or shoot the opposition) and are less likely to learn of problems or be willing to act to solve them. If achieving sustainability requires an ongoing process of trial, error, and learning applied to further trials, as suggested by the principles of adaptive management, then the system must provide ways for detecting such errors and acting on them.

Nevertheless, despite its appeal, it isn't clear that democracy is better for sustainability, and we cannot say with certainty that it is. It is possible (barely) that an enlightened despot might come to power somewhere and put his country on the road to sustainability by edict. Apart from its philosophical and ethical advantages over other political systems, however, we believe democracy—even in the compromised forms in which it is practiced today—*is* better for sustainability. Also, a deepened democracy (we might say a *true* democracy) likely would be the most conducive of all forms of governance to sustainability. In this chapter, we will explore a form of democratic governance that offers real hope of achieving a just and sustainable society.

It is fair to ask why a change to a deeper form of democracy is needed. Why doesn't our current system offer the same hope? The answer, at least in part, is because it is not really a democracy at all. We have been using "democracy" to mean the system of government in place in the United States and in most other developed nations. Strictly speaking, that is inaccurate. High-school civics courses make the simple distinction between democratic and republican forms of government. In democracies, rule is by the people, and in republics, rule is delegated to an elected body of representatives. The United States and the rest of the developed nations are republics, and to one extent or another they all suffer from the same malaise of widespread citizen disengagement and rule by professional politicians, who are courted and swayed by interest groups, lobbyists, and money. The head governs, but the body politic is a corpse. "Most Americans," writes journalist Robert Samuelson, "consider the freedom from politics to be part of their well-being."[4]

So we are free to pursue happiness as consumers. But this arrangement seriously limits the capacity of "democratic" political systems to carry their citizens toward sustainability. Samuelson's insight points straight to the contradiction at the heart of the industrial democracies. It is partly the freedom and productivity of democratic systems that has allowed so many people to do so well economically,[5] yet (at least in the United States) engagement in those political systems is at historic lows, as judged by such standard measures as voter turnout. But democracy by definition is a system that requires participation. Unless we are willing, explicitly or by default, to cede all power to a political/bureaucratic/technical elite, the only issue is how much participation.

It also appears increasingly obvious that support for exploring sustainability options, except among a small minority of concerned citizens, is weak. The issue itself hovers well below consciousness for the general public. Mention "sustainability" to most people, even intelligent and well-read ones, and the nearly inevitable response is, "Of what?" Poll data are disturbing as well. They confirm that Americans, like citizens in most countries, like to think of themselves as environmentalists. But consider two paired surveys of about seven hundred Americans each, one conducted before the White House conference on global climate change in October 1997 and the second in early 1998 after the United States had signed the Kyoto Protocol on reducing greenhouse gas emissions.[6] The heavy media coverage generated by these two events triggered little change in peoples' opinions about global warming. The changes that did occur had to do mainly with what should be done, and by whom. They were not heartening:

> Ninety-one percent of people in December–February [the second poll] said the U.S. government should limit air pollution by U.S. businesses, up somewhat from 88 percent in September–October [the first poll]. Likewise, 80 percent of people in December–February said the United States should require air pollution reductions from countries to which it gives foreign aid, up from 71 percent in September–Octo-

ber. Yet *fewer* people were willing to pay higher utility bills to reduce air pollution: 72 percent in December–February, as compared with 77 percent in September–October.

In the United States, 19 percent of all emissions of carbon dioxide, the chief greenhouse gas,[7] is attributable to the residential sector (houses and apartments using energy derived from natural gas or coal for heating in winter and cooling in summer). Another 32 percent comes from the transportation sector,[8] essentially meaning cars and trucks, including the 2- and 3-ton 4x4s sitting in suburban driveways all over America, ready at the turn of a key to whisk their owners to the mall at 13 miles per gallon. Thus, more than half of U.S. carbon dioxide emissions are linked directly to individual decisions made by consumers. As for those foreign-aid–receiving countries: of course they should cut their emissions. We want our money's worth! But the United States is the world's largest emitter of greenhouse gases; its share is almost as large as those of the next three countries (China, Russia, and Japan) combined.[9] U.S. per capita carbon dioxide emissions are many times higher than such emissions in most countries and are exceeded only in Norway, Kuwait, and United Arab Emirates—all of them heavily dependent on oil and gas production.[10]

Or consider this anecdote. People are flocking to the American Southwest, part of the desert sunbelt, but taking along the suburban values of the lusher regions they leave behind. They still want well-watered lawns and verdant golf courses, and they often do not grasp the need for increasingly stringent controls on water usage as ballooning populations strain scarce water supplies. A Las Vegas Valley Water District investigator told of confronting a homeowner about an illegal sprinkler. "He got so angry," the investigator said, "he poked me in the chest and he said, 'Man, with all these new rules, you people are trying to turn this place into a desert.'"[11]

Even well-intentioned people imagine that they can somehow phone in their contributions to a sustainable society. American popular environmentalism is a mile wide and an inch deep, and our grasp of the contribution we make to environmental problems is generally superficial. It seems obvious that little further progress toward sustainability—at least, little conscious, intentional progress—can be made unless that grasp

Figure 5.1. The paradox of American environmentalism
Reprinted with permission of King Features Syndicate

becomes much firmer. This is perhaps the most important reason for seeking a more genuinely democratic political system, and we will return to it in a moment.

Why We Need a New Politics

First, we turn to two other practical arguments, beginning with the emergence of "postnormal" science and the need to broaden the stakeholder base regarding issues of sustainability.

Some ecological economists believe that science has entered a new phase in which the nature of the problems it faces, and therefore its fundamental function, have changed.[12] In the old era of "normal" science, which was based heavily on mechanistic views of the universe, science functioned to improve production and was the servant of economic growth. The problems science faced were in the nature of puzzles. Science's challenge was simply to solve them. What is the best way to increase wheat yields? Breed a new strain that thrives on less water. How can aircraft be made faster? With piston engines at the peak of their development and thus at a dead end, invent turbojets. How can we make lots of cheap electricity and thereby end our dependence on imported sources of energy? Adapt nuclear reactors used in submarines to generate power for the grid. All these scientific enterprises seemed like good, straightforward ideas at the time. Right or wrong, science and technology in general were seen to be value free, though of course they could be put to good or bad uses.

This era is passing away. Science can no longer simply solve puzzles "in ignorance of the wider methodological, societal, and ethical issues raised by the activity and its results."[13] One reason is uncertainty. The most important problems science must now address, including those related to sustainability, can no longer be seen as reducible to mechanistic explanation, because they frequently are embedded in complex systems with inherently unpredictable properties that involve global events or trends and have global implications that affect every person, indeed every creature, on the planet.

The unpredictability stems from the fact that ecosystems, including the global ecosystem, are not merely complicated, they are *emergently complex*. A clock is complicated in the sense that it has many pieces linked by intricate interactions. But no matter how complicated, it either runs or it doesn't. Its mechanisms are linear and its operation straightforward. It may slow down as the mainspring uncoils or as the bearings gum up, but it won't suddenly begin to run backwards. A clock also has no resiliency; removing almost any part will cause it to stop.

Ecosystems, too, have many parts interacting intricately, and like clocks they run on inputs of energy from outside (i.e., the sun). But they show important additional dimensions of behavior. Ecosystems are dynamic: they grow and change continuously. Removing a part may have no important effect at all—or, if the part is crucial to the ecosystem's function, removing it may cause the system to become something quite different. On the other hand, ecosystems can bounce back from many insults.

Above all, ecosystems display *emergence*. Emergence means that a system can show properties or behaviors that cannot be predicted from breaking it down into its component parts and understanding them reductively. Water, for example, is pretty simple in molecular terms: one oxygen atom with two hydrogen atoms "stuck to it like Mickey Mouse ears."[14] The behavior of a single molecule can be described in terms of well-understood equations. Those equations, however, give little hint that combining billions of water molecules produces a substance that transforms from a solid to a liquid at 32 degrees Fahrenheit and thence to a gas at 212 degrees, or that can inspire countless poets with its magical beauty. Likewise, life itself displays emergent properties: biological processes can be described thoroughly in terms of biochemistry, but those biochemical laws do not predict whalesong, the *Mona Lisa,* or Play-doh.

In emergently complex ecosystems, new species appear and others die out. The balance among the pieces may change, and the nature of their interactions shift. An ecosystem may begin to show new dynamics and characteristics. Surprises are normal, and the unexpected, particularly in response to perturbation, happens all the time.

Although humans are constantly being surprised by nature, it is not merely due to lack of information. It is not necessarily true that if we only knew more about the preconditions, we would be able to predict what was going to happen. The emergent complexity of ecosystems means that *their behavior is often intrinsically indeterminate and unpredictable.* Because human beings are totally immersed in and part of ecosystems, we are subject to that indeterminacy. We also contribute to it. "Any natural system that is of interest to us has properties that affect our welfare."[15] We must live with unavoidable uncertainties.

This relentless uncertainty complicates the role of science. In the "normal" phase, science was seen as providing solutions to puzzles and assigning probabilities to the possible outcomes of policy decisions. Scientists, as the experts, could supply the answers, and the stakes riding on any given decision could be kept relatively low. You could figure out how dangerous an action was likely to be and then decide how, or even whether, to do it, or what limits to set on it. The whole discipline of risk assessment is based on this assumption. The idea was that scientists could determine quantitatively how likely it was that a new chemical, say, would be toxic or carcinogenic, and how rapidly and completely the environment would absorb and neutralize it. Those estimates would make it possible to set a "safe" level of exposure and thus establish standards for emissions. After more than twenty years' effort, however, we see that risk assessment clearly often fails in the face of the complexity of the tasks it attempts. Scientists disagree on how to assess damage to the human immune and nervous systems. Assessing the damage done by multiple exposures to thousands of chemicals with unknown interactive effects is impossible. Equally impossible is determining how individuals who differ wildly in physical health, obesity, diet, and dozens of other factors might react to a given chemical. For these and other reasons, risk assessment winds up being more art than science.[16]

When the situation is profoundly uncertain, the decision stakes—the potential impacts of harmful outcomes if the wrong decision is made—go up. Sustainability issues are global in scale, staggeringly complex and interactive, poorly understood, and riddled with uncertainties. Because of the inherent unpredictability of the natural systems that support us, the possible outcomes of environmental policy decisions are highly uncertain and therefore highly risky. Policymakers seek solid facts upon which to base predictions and policies, but there aren't many of those. In the face of such complexity and uncertainty, all science can offer is probabilities and forecasts based on untestable models. Consequently, "Uncertainty and ignorance can no longer be expected to be conquered; instead, they must be managed for the common good."[17] Where once science was a source of authoritative advice on how to extract and use resources more rapidly, with the goal of steadily widening the stream of new wealth, now science must become just one partner in a broad-based decision-making process that involves anybody with a stake in the outcome. With global environmental problems and issues of global carrying capacity, essentially everyone is a stakeholder.

Postnormal science, the only scientific perspective truly applicable when "facts are uncertain, values are in dispute, stakes are high, and decisions are urgent,"[18]—the very situation we face today—therefore not only cannot expunge politics from the sustainability question but must become subordinate to politics in addressing sustainability. This does not mean that science has nothing to contribute to the debate about sustainability, but it does throw open the forum to others acting in a critical role. Under the old "normal" scientific paradigm, critics of a scientific position mainly consisted of "out-group" scientists who thought existing theory could no longer be stretched to explain all the data adequately. Today, the problems we face are routinely unsusceptible to explanation by the experts, since nobody really has final answers when uncertainty and complexity rule. Scientists are no longer the only "legitimate" participants in the decision-making process (if indeed they ever were). Hence, the "peer community" must be widened to include those with differing expertise of their own to offer, as well as to those with a significant stake in the outcome of the deliberations:

> When problems do not have neat solutions, when the phenomena themselves are ambiguous, when all mathematical techniques are open to methodological criticism, then the debates on quality [of decision outcomes] are not enhanced by the exclusion of all but the academic or official experts. . . . The democratization of science in this respect is therefore not a matter of benevolence by the established groups. Rather it is . . . the creation of a system which in spite of its inefficiencies is the most effective means for avoiding the disasters that in our environmental affairs . . . result from the prolonged stifling of criticism.[19]

That's one reason for broadening the stakeholder base. Another is the existence of incommensurabilities, the things that can't be compared easily or at all. "If emissions of CO_2 diminish while those of SO_2 rise, how can one decide whether the state of the environment has improved, got worse, or stayed the same?" Many things that involve trade-offs cannot all be reduced to the same kind of index, allowing a simple numerical comparison. Moreover, people often attach priceless values to aspects of their environment, rendering quantification meaningless.[20] Value choices must then be made, and that is the domain of everyone, not just experts.

A third reason is local knowledge. Not only does the stake that nonexperts or local people have in the outcome of an issue rest on solid ethical grounds, but "Those whose lives and livelihoods depend on the solution of the problems will have a keen awareness of how general principles are realized in their 'back yards.'"[21] Farmers in the north of England, for example, reportedly have shown a better grasp than the experts of how radioactive contaminants from the nearby Sellafield nuclear fuel reprocessing plant actually behave in the thin soils covering the moors where they graze their sheep.

In the end, these factors point not to a scientific process merely shaped by social and political forces but to a social and political process aided (but not controlled) by scientific inputs. That makes the process and responsiveness of politics more important than ever:

> There is no question of a complete specification of possible outcomes of human actions in their ecological setting. The question of what will constitute a satisfactory collective decision-making procedure will not depend on eliminating the uncertainties and dangers, nor masking over them. It will have more to do with how the risks are identified, discussed, and shared along the way—that is, with how agreements and compromises are reached, and how fears and losses are acknowledged and respected.[22]

So much for the broadening stakeholder base, the first practical reason for a more engaging politics. The second reason has to do with the process of adaptive management touched on in chapter 2. The complexity and uncertainty of the world and the things we are doing to it by our economic activity pose great risks. If we wish to pursue sustainability, the prudent response to these risks is to systematically treat what we do to the world as an experiment and take careful note of what happens. That is the heart of adaptive management. When the information thus gathered is processed and applied in a context of bounded conflict (politics), social learning occurs.[23] Deeper engagement by ordinary citizens in this process is important in at least three ways: it gives legitimacy to the conflict by broadening the scope of representation of different points of view; it helps ensure that the system is maximally sensitive to the detection of problems; and it can provide the kind of political support for the long-term science necessary to develop and refine ways of managing ecosystems sustainably.

Is this ideal attainable? In *Compass and Gyroscope,* his insightful book about adaptive management applied in the Columbia River basin, Kai Lee recounts John Dewey's 1926 call for "an experimental politics" as a solution to "the puzzle of how a mass society could deal with questions that necessarily required expert knowledge to articulate or decide." Dewey argued that the complexity of mass industrial society encouraged most people to cling to comforting dogmas (such as communism and anti-communism) that offered to make sense of the world even if they painted false or distorted pictures of it. As an alternative to this "battle of dogmas,"

Dewey's experimental politics assigned citizens and experts distinct roles according to their natural strengths: citizens were uniquely suited to recognizing and identifying social problems needing attention, setting the agenda, and staging the debate. Experts would act as "teachers and interpreters" to make sense of the technical complexities of the issues—to inform policy but not to make it.[24]

This vision is very much in tune with the politics of engagement we are proposing. However, Professor Lee throws cold water on it: "Social research since the 1920s has made Dewey's confidence about the rehabilitation of democracy seem less plausible."[25] Lee argues that people do not seem up to the job of agenda-setting and debating; that the experts cannot be trusted to merely offer objective analysis but instead become hired guns for any party with the money to buy their testimony; and that "public attention has displayed little durability and stability in the face of subtle and complicated scientific problems, a long public agenda, and intense competition to attend to many other things in life besides affairs of state."[26] The result, he says, is that citizens play no important role. It is "policy subsystems" (institutions), not citizens or elected officials, that pay long-term attention to problems. Experts do not act as teachers of the citizenry. "Learning occurs, ideas matter; citizens and democracy in a direct sense do not."[27]

All this is a more or less true and accurate description of current circumstances: the system *is* adversarial, partisan, bureaucratic, dominated by interest groups, and run by an elite of politicians and technocrats. What we are suggesting is the potential superiority of a different set of circumstances. If mass society is complex and daunting, perhaps it would benefit by a certain degree of de-massification—that is, by increased focus on localities. If citizenship as a public ideal is weak, perhaps it is because our current system places little value on any feature of it except voting (frequently a token act; as a restroom graffito has it, if voting could really change things, it would be illegal). If people seem incompetent to set agendas, weigh the opinions of experts, and debate policy options (which they can be competent to do, as evidence in chapter 6 will show), perhaps it is because the system has never been structured to invite or require them to do so, nor set up to educate people into citizenship. Dewey's vision suffers from not having been tried. Yet it offers potentially compelling benefits, in the legitimation of bounded conflict and the sensitivity to detection of problems mentioned above.

This brings us to the third and, with respect to sustainability, perhaps most important reason for a politics of engagement. Although most people play several roles in their lives, "The dominant preference orientation for most decision makers continues to be that of the consumer."[28] One consequence is that, as has often been said of economists, we act as if we know the price of everything and the value of nothing. The failures of market systems to reveal the full costs of goods and services shield consumers from the environmental effects of their economic behavior. While railing, a few paragraphs ago, against the fuel inefficiency and pollution of the massive 4x4s currently so popular among American consumers, we did not mention the additional, "upstream" burdens their manufacture places on the environment, such as the energy expended and the mining and drilling wastes produced to make the steel, aluminum, glass, rubber, and plastic. Few of us could enumerate these burdens if asked and fewer still bear them in mind when visiting auto showrooms. Moreover, in our isolated role as consumers, we are poorly equipped and motivated to question the whole rationale behind a transportation system that encourages and subsidizes the growth of vehicular traffic, the perpetual expansion of roads to relieve congestion, and the catch-up expansion of traffic to fill the new roads—a vicious traffic circle if ever there was one.

Similar criticisms could be made of most consumer goods. The behavior of consumers makes sense when viewed from the logic of consumerism; but too few incentives exist to adopt another logic—such as that of citizen. Motorists do not make decisions about transportation systems. They complain when the potholes aren't filled and grumble about unsynchronized traffic lights and rush-hour gridlock, but they do not sit down to discuss the general issue of mobility in their communities and how best the community's transportation needs might be met. That is left to anonymous traffic engineers and urban planners with quite different motivations and stakes in the policy outcome. This arrangement reinforces the general belief that people need not exercise power or responsibility because someone else will. Leave the job to the professionals—the legislature, the city council, the courts, the bureaucracy—who then become convenient scapegoats when things do not work.

Undeniably, people are more than consumers. Because of the various legitimate roles people fill, they will have various equally legitimate preference orientations. These orientations will sometimes be at odds; for instance, one's interests as a consumer in the availability of sophisticated art films for adults may conflict with one's interests as a parent in guarding access to such films by the underaged. The point is that the citizen preference orientation is currently attenuated to the point of invisibility. Yet strengthening it would ineluctably bring people face-to-face with the problems of governance, including those of sustainability. Citizens brought into stark confrontation with the problems of governing their communities through hands-on participation on commissions, juries, zoning and school boards, PTAs, and other structures of civic decision making (including, we can hope, neighborhood assemblies; see the section titled "How Strong Democracy Might Actually Work," later in this chapter) would be educated in the sources of community troubles, in the origins of their way of life, and in the trade-offs that must be accepted in any collective choice. With regard to sustainability issues in particular, self-governing citizens would more likely learn the ecological costs of their community's lifestyle and socioeconomic character. Political scientist Lamont Hempel puts it this way:

> Global environmental protection begins at the community and bioregional level—the level where complex living systems are most interdependent and vulnerable. Local watersheds, ecosystems and microclimatic conditions are among the primary objects of bioregional protection, and their alteration by human activities is much easier to understand from the vantage point of local communities than from the macro perspective of global ecology.[29]

By having to involve themselves directly in the resolution of conflicts, engaged citizens would more clearly see how the wish for green space contests with the wish for suburban development; how an automobile-addicted culture enables pollution, sprawl, and the strip malls so many people say they hate; how extravagant use of fertilizers and pesticides on croplands and suburban yards contaminates water supplies; and so on.

A politics of engagement would help transform the popular view of politics as "the conduct of public affairs for private advantage," as Ambrose Bierce once commented, with characteristic cynicism. An engaging politics would short-circuit our "procedural republic," the fantasyland of citizens deliberately separated from one another by a structure of delegational authorities (representatives, officials, judges) to whom appeal is made for benefits and favors at the expense of all the other supplicants. An engaging politics would help to overcome the disconnection between the governing and the governed, to reduce the sense that "they" do things to "us." Ideally, "they" would be "us"—and we would be paying more attention to our collective affairs because our collective decisions would matter.

One View of Strong Democracy

We mean, again, to be mostly noncommittal about the particular form or forms sustainable societies might ultimately take. Whatever that form is, though, a prerequisite is a political system that allows the various visions and the processes by which they are to be pursued, to be discussed, and worked through in an adequate way. "Adequate" here means that enough people are involved for long enough that whatever choice is made can be said to be conscious, comprehensive (widely based), comprehensible, and comprehended. The politics of engagement we have been describing could be the framework for that process of discussion, working through, and self- and mutual education.

One compelling model for a politics of engagement is strong democracy, the "road not taken" in American political history. More than two hundred years ago, the framers of the Constitution spent a hot summer in Philadelphia debating the virtues of "republicanism"[30] and federalist forms of government in their struggle to define the shape of civil society in postrevolutionary America. As recounted by Daniel Kemmis,[31] the delegates to the 1787 Constitutional Convention were wrestling with one overriding question, prompted by the recent uprising of Daniel Shays and other Massachusetts farmers who were deeply indebted to city-based lenders. The question on everyone's minds was whether the political system should be arranged so that citizens worked through such disputes themselves, or so that government institutions intervened, like parents separating squabbling children. This was the famous argument between Thomas Jefferson and the republicans on the one hand, and James Madison, Alexander Hamilton, and the federalists on the other. In effect, the republicans wanted to bring citizens together, the federalists to keep them apart.

Jefferson believed fervently in the prospects of educating people into citizenship, their capacity to thereby perceive the common good (*res publica,* the public thing), and thus to largely handle their own affairs and forge public policy. He dismissed Shays' Rebellion as an isolated incident. He believed in the potential for engagement and interaction, an approach that, in Kemmis's words, "assumed that citizens were presented with many opportunities and much encouragement to rise above narrow self-centeredness. Only if citizens were, in various contexts, putting themselves in one another's shoes could they be expected to identify with and act upon a personally perceived vision of the common good."[32] The development of that capacity was the object of education into citizenship.

The federalists, in contrast, saw Shays' uprising and other ongoing conflicts as evidence of the inability of ordinary citizens to control their passions and work out their differences, and as a profound threat to stability and order. The colonies' newly won independence must have seemed precarious, and there was no time to wait for people to be educated into citizenship, even if they could be. The federalists' first alternative to engagement, which came to be incorporated into the Constitution, was the elaborate system of checks, balances, and other governmental machinery intended to keep citizens apart: the "procedural republic." The second alternative was the frontier, which acted as a safety valve by allowing disaffection and conflict to be mitigated by migration rather than politics. "Republicans," Kemmis writes, "believed that public life was essentially a matter of the common choosing and willing of a common world—the 'common unity' (or community), the 'public thing' (or republic). The federalists argued that it was possible—in fact it was preferable—to carry on the most important public tasks without any such common willing of a common world." Not accidentally, "the federalist plan of government was exactly analogous to Adam Smith's invisible hand, which wrought the highest good in the market even though none of the actors were seeking anything beyond their own individual interest. Smith intro-

duced the concept of the invisible hand into economics in 1776. Twelve
years later, Madison introduced it into politics."[33]

Though the federalists carried the day at the Constitutional Convention, Jeffersonian republicanism nevertheless survived as an active political philosophy for many years and grew in strength until late in the nineteenth century. Its finest moment came in 1896, when the agrarian
populist movement nearly succeeded in putting William Jennings Bryan
in the White House. William McKinley won the election, however, beating
Bryan and the populists with a politically innovative blend of industrial
money, Madison Avenue smarts, and public relations:

> The populists had built their power by teaching people new
> methods of cooperation, both in economics and in politics.
> In a variety of ways, this movement had brought people
> face-to-face with each other so that they could work out,
> among themselves, the possibilities of a better way of life.
> People within the movement had learned to depend upon
> each other, to trust each other, to educate each other. In all
> of these ways, the populist movement depended upon pre-
> cisely those "civic virtues" and republican principles of face-
> to-face politics which had been so important to Americans
> from John Winthrop to Thomas Jefferson. But in the end,
> the Populists were defeated by a system that relied, not upon
> those face-to-face dealings, but upon the highly impersonal
> methods of mass communication.[34]

This system—in which a Senate campaign costs millions and a presidential campaign tens of millions—is the one still in place today. Tied
with these structural effects, the 1896 election also had a far-reaching psychological impact. "The fundamental faith of ordinary people in their
ability to govern themselves had been lastingly diminished," Kemmis
writes, citing historian Lawrence Goodwyn to illustrate the depth of the
damage done:

> Older aspirations—dreams of achieving a civic culture
> grounded in generous social relations and in a celebration of
> the vitality of human cooperation and the diversity of

> human aspiration itself—have come to seem so out of place
> in the twentieth century societies of progress that the mere
> recitation of such longings, however authentic they have
> always been, now constitutes a social embarrassment.[35]

People naturally shrink from such embarrassment and in any case need little urging to retreat to the gilded cages consumerism has built for them. But if ecological constraints on economic growth are real, then consumerism is just one more rapidly contracting frontier. What happens when it, too, is closed? The elaborate procedural republic we have built in the United States is nearly strangling on regulation and litigiousness already. How much more can the system handle?

Perhaps we should investigate the practice of handling our differences in another way. No one knows how things might have turned out if the Framers had opted for a political system that stressed citizen engagement rather than separation, but we can explore the idea of engagement in a modern light. Although many political theorists have written about strong democratic governance, perhaps the defining vision is that of Rutgers University professor Benjamin Barber.[36] Barber has eloquently developed and defended his ideas about strong democracy in his book by the same name and in subsequent writings. We can only sketch an outline of this complex vision here, and we have edited his vision somewhat to suit what we see as the interests of building a sustainable society. We begin with the definition of strong democratic politics offered in chapter 1:

> Politics in the participatory mode where conflict is resolved
> in the absence of an independent ground through a partici-
> patory process of ongoing, proximate self-legislation and
> the creation of a political community capable of transform-
> ing dependent, private individuals into free citizens and par-
> tial and private interests into public goods.[37]

The first and most important element is thus participation. Including as many people as possible in exercising power is intrinsic to strong democracy. In contrast, the classically liberal democratic system now in place in the United States, which Barber calls "thin democracy," is more or less designed to exclude most people, because conflict is seen as "a problem created by political interaction rather than the condition that gives rise to politics."[38] Classic liberal democrats reject participation because they profoundly distrust power and its capacity to infringe on private interests and freedoms. There is a widespread fear that too much participation will merely open the system's door to the "wrong" kind of people and, by empowering them, give them a wider stage upon which to express their various shortcomings as human beings.[39] Liberal democracy views people in Adam Smithian terms, as social atoms, "whose every step into social relations, whose every foray into the world of Others, cries out for an apology, a legitimation, a justification. . . . Politics is prudence in the service of *Homo economicus*—the solitary seeker of material happiness and bodily security."[40] Liberal democracy holds fast to beliefs in "the fundamental inability of the human beast to live at close quarters with members of its own species"[41] and in the idea that the only connection between people is the self-interested bargain.

Liberal democracy, like the capitalism it is usually mated to, also assumes that human nature is essentially unchangeable. Strong democracy, on the other hand, asserts that human nature is the product of culture and history: "It is from common rather than individual consciousness—from generations of communal labor and not the passing whimsies of individuals—that the enduring features of human identity are born. We are above all creatures of time, defined by a history that we make together."[42] Humans living together possess what theologian Reinhold Niebuhr called "the primordial character of human community"[43]—the outcome of millions of years of evolution in social groups.

Strong democracy is a superior response to the challenges of the essential political condition, summed up in this question (which is also the key question of sustainability): "What shall we do when something has to be done that affects us all, we wish to be reasonable, yet we disagree on means and ends and are without independent grounds for making the choice?"[44] The pursuit of sustainability clearly evokes all these conditions. The global context of sustainability means that it affects everyone. Reasonableness is not only ethically preferable to mulishness or violence in exploring the issues, it may be the only socially sustainable response. We certainly disagree about means and ends, and often about whether sustainability as a complex of issues is even important. And we lack an *agreed-upon* common independent ground, or source of objective value—for example, a particular notion of God and an interpretation of God's will—to refer to in seeking guidance on how to act.

This last point deserves elaboration. Strong democracy is very far from arguing that there is no such thing as objective value. In fact, there is no point in preferring sustainability, and maintaining that strong democracy is most conducive to it, unless it can be demonstrated that a sustainable society is better than one that isn't. Such an argument has to be made on the basis of some higher value: that the biosphere is God's creation and therefore deserves respect and care, or because sustainability means less suffering in the long run, or some other line of thinking. But as a form of political practice, strong democracy does not depend on agreement on a form of objective value; it instead picks up where agreement on objective value is absent: "Where there is certain knowledge, true science, or absolute right, there is no conflict that cannot be resolved by reference to the unity of truth, and thus there is no necessity for politics. Politics concerns itself only with those realms where truth is not—or not *yet*—known."[45] Strong democracy is thus an especially apt response to sustainability issues—in which "facts are uncertain, values in dispute, stakes high, and decisions urgent"—because the issues cannot be addressed collectively by appeal to a "higher power" or universal system of beliefs and guidelines. While many or most of the participants may hold strongly to various forms of objective value, there are none that are held in common. Nor can science supply the answers in any final way, only successive and useful bits of feedback about the effects of our actions on the world. Strong democracy is also appropriate to pluralistic societies like the United States, where the wildly varying spectrum of political and cultural orientations includes fundamentalists of every stripe, from those who believe in One Truth, whether it is God or scientific nihilism, to those who believe (with Gertrude Stein) that "There ain't no answer. There ain't going to be an answer. There never has been an answer. *That's* the answer."[46]

Barber lists seven attributes of the political condition: action, publicness, necessity, choice, reasonableness, conflict, and absence of an independent ground. Strong democracy, he suggests, is the best response to these conditions:

ACTION. Politics means action by humans that makes a difference in the world. Politics is something people do, not merely have or watch. Citizens are therefore actors in the political drama, not just spectators hiding behind the footlights of a play where politics is the exclusive province of politicians, lobbyists, and bureaucrats. The political action of citizens under strong democracy is marked by common deliberation, common decision, and common work.

PUBLICNESS. Political action is public action; it has to do with the *we*. Private actions that do not have public effects are not the business of politics. One of the key questions of politics, therefore, is when do private acts become public? The purpose of politics is to create an ongoing process by which that question can be continually answered as society's ideas of what is the public (and therefore what is the community) evolve. Strong democracy creates a public that can debate such questions; in fact, it sees as one of its chief tasks the creation of a civic community by and for such deliberation.

NECESSITY. Political action means necessary action in the world, where events have lives of their own and to refuse to act is also to act. In a participatory democratic system, citizens are embedded in and engaged with the world in a way that makes the necessity and inevitability of political action, and responsibility for that action, inescapable.

CHOICE. The need for action implies the need to choose which action. Those who do the choosing must be of a particular type: "Just as we would not understand a sleepwalker to be a human agent or a hysteric to be a human actor, so a rabble is not an electorate and a mob is not a citizenry. If action is to be political, it must ensue from forethought and deliberation, from free and conscious choice."[47] A citizen is someone who exercises that free and conscious choice, that is, who is autonomous. True participation presupposes autonomy; any other kind of participation is a sham. The political community conceived of in strong democracy is both made up of autonomous citizens and seeks to cultivate them. The type of liberal democracy we currently live in takes a narrower view of what is expected of citizens and what they can be trusted to do.

REASONABLENESS. The elusiveness of a commonly agreed-upon independent ground for judging choices suggests that we have to find a middle way between choices dictated by Truth (which we may or may not agree on, even if each of us believes he or she is working toward it) and choices that are made or argued for merely on the basis of the simple assertion of individual tastes. The choices should be reasonable, in the sense of practical, not arbitrary and not forced on people. Reasonableness thus is a kind of social glue that allows choices to be made without creating irrevocable disaffection within the community. Reasonableness acknowledges that "while this seems the best choice, you have some valuable points and I may be wrong." Strong democracy cultivates reasonableness by emphasizing the importance of talk and listening, about which more below.

CONFLICT. Conflict is inevitable in human society. Politics arises as a response to conflict; as we saw in chapter 3, it is only in utopias artificially made free of conflict that there is no need for politics. Societies differ in how they decide to handle conflict politically. (When conflict is settled through violence, society can be said to have collapsed to one extent or another.) Strong democracy does not demand consensus, but seeks to "transform conflict into cooperation through citizen participation, public deliberation, and civic education."[48]

ABSENCE OF AN INDEPENDENT GROUND. People seek certainty despite its perpetual elusiveness, no less in politics than in other areas of life. Political theorists and policymakers seek certainty in the form of a fundamental guiding principle that exists outside politics, to which politics can then refer in resolving conflicts. Life would be simpler if we all agreed that the Old Testament, or Plato, or the National Academy of Sciences, or *The Nation* should be our ultimate guide. But in any diverse, pluralistic society, complete agreement on such an overriding principle is unlikely. In the United States and other countries, the assumption by many groups that there ought to be such a principle—theirs—itself creates a great deal of conflict. Strong democracy, while creating a forum and a process for mutually pursuing Truth, assumes that complete agreement on it in the form of a guiding principle is improbable: "The procedures of self-legislation and community-building on which it relies are self-contained and self-correcting and thus are genuinely independent of external norms, prepolitical truths, or natural rights."[49] Strong democracy acknowledges that participants inevitably bring values to the political arena, whether they are a thoughtless, unexamined mishmash of bromides or a coherent philosophy rigorously and painfully arrived at. In either case, strong democracy puts values to the test in a public process of talking, listening, and reasoning together:

> The autonomy of the democratic process under strong democracy equalizes value inputs. It gives to each individual's convictions and beliefs an equal starting place and associates legitimacy with what happens to convictions and beliefs in the course of public talk and action rather than with their prior epistemological status.... Politics in the participatory mode does not choose between or merely ratify values whose legitimacy is a matter of prior record. It makes preferences and opinions earn legitimacy by forcing them to run the gauntlet of public deliberation and public judgment.[50]

Everyone is, of course, free to argue on the basis of personal preference: "Blue is better than green because I like it more." That position may even be called a piece of the truth—but it is a very small, trivial piece, not of great significance to the welfare of the larger community, and unlikely to carry much weight. It does not offer the community a useful new way to approach the truth and thus fails the "So what?" test. Anyone making such arguments on more important matters will rapidly find the community attending instead to others with more thoughtful views. The process, part of education into citizenship, thus gives participants the opportunity to reexamine their beliefs, and the extent to which those beliefs may represent progress toward perceiving the truth more clearly, in light of the feedback provided by a community of others engaged in the same quest.

Barber compares strong democracy to four other theoretical types of democratic systems (see table 5.1), none of which exists in pure form. In none of the cases is there supposed to be an independent ground, but in all, some conceptual force fills that role surreptitiously. There are three of

Table 5.1. Democratic Regimes (Ideal Types)

Regime Form	Political Mode	Value	Institutional Bias	Citizen Posture	Government Posture	"Independent Ground" Disguised as
REPRESENTATIVE DEMOCRACY						
Authoritative	authority (power/wisdom)	order	executive	deferential unified	centralized active	noblesse oblige wisdom
Juridical	arbitration and adjudication	right	judicial	deferential fragmented	centralized limited	natural right higher law
Pluralist	bargaining and exchange	liberty	legislative	active fragmented	decentralized active	invisible hand natural equality market rules
DIRECT DEMOCRACY						
Unitary	consensus	unity	symbolic	active unified	centralized active	the collective the general will
Strong	participation	activity	populist	active centralizing	decentralizing active	(no independent ground)

the "thin," representative type (authoritative, juridical, and pluralist), and two direct types (unitary and strong).

Authoritative democracy is marked by an activist, centralized structure with a strong executive (such as a Franklin D. Roosevelt) working with and through a governing elite. It values order and security. The citizenry defers to the elite and is relatively unified, but the unity is defined by the governing elite's perception of the masses' interests. Participation is limited to elections. The independent ground is disguised as the wisdom and virtue of the political elites.

In *juridical* democracy, the first value is right and the protection of "rights," which it seeks to uphold by means of an activist representative judicial elite. This elite enforces the protection of rights by arbitration and legal rulings. Sometimes its activism seems to stray over into the legislative realm. Citizens are again deferential. The emphasis on rights, however, tends to set citizens against each other. The independent ground is disguised as "higher" law and constitutional norms. This strain of democracy will seem familiar to Americans used to seeing people demanding new "rights" every day and the courts only too happy to rule on their right to enjoy those rights.

The *pluralist* democratic model stresses freedom and seeks to ensure it by relying on bargaining and exchange by individuals and organized groups in political markets where every party pursues its own interests. The whole system is held together by a social contract that defines the rules for bargaining and enforces the results. By definition, the citizenry is more active than in the previous two types, and the government is less centralized because power is more broadly spread out. Barber offers the laissez-faire conditions in nineteenth-century England as an example, though in many ways late-twentieth-century America doesn't seem far-fetched either. The hidden independent ground here is the market itself, the notion that the invisible hand creates the public good through the restless seeking of individual self-interest, whether economically or politically.

Among the two direct types of democracy, *unitary* democracy aims, as the name implies, for consensus and unity. This may depend on a unifying characteristic, such as race or ethnicity (real or presumed) that gives the populace a measure of homogeneity. The government is centralized and active, and seeks to embody the common will symbolically (e.g., the Third Reich). Citizens do not so much take part in the greater whole as become subsumed by it. The community tends toward conformity and sometimes toward coercion. In large settings, this tendency can become exaggerated to the point of complete corruption of the democratic impulse. The "collective will" as interpreted by the government functions as the independent ground.

Barber argues that all the above forms of democracy suffer from two defects. First, except for unitary democracy, they rely on representation in one form or another, and "representation is incompatible with freedom because it delegates and thus alienates political will at the cost of genuine self-government and autonomy. . . . Men and women who are not directly responsible through common deliberation, common decision, and common action for the policies that determine their common lives are not really free at all, however much they may enjoy security, private rights, and freedom from interference."[51] When people delegate their power and responsibility for governance, they do not lend them but give them away. Unitary democracy nominally does not delegate but in effect gives away power and responsibility to the collectivity, which is represented by the government.

Second, while the representative forms of democracy purport to avoid independent grounds, in fact they all sneak them back into the process in different guises. That is, they assume some prepolitical, philosophical idea or foundation to which people can refer during the political process—when in fact the realized forms of those ideas (freedom, rights, etc.) are the *outcome* of politics, not its antecedents. The deference to an extrapolitical truth chops politics off at the knees by replacing the interactive reflection and deliberation of citizens with abstractions. The result is a politics that resembles a movie set or one of those flyblown western cow towns of yore, with the two-story facades on one-story buildings: there is much less there than meets the eye. "Politics has become what politicians do; what citizens do (when they do anything) is vote for the politicians. . . . Liberal and representative modes of democracy make politics an activity of specialists and experts whose only distinctive qualification, however, turns out to be simply that they engage in politics."[52]

This brings us back to strong democracy. To simplify the definition offered earlier, strong democracy means that people—citizens—govern themselves to the greatest extent possible rather than delegate their power and responsibility to representatives acting in their names. Strong democracy does not mean politics as a way of life, as an all-consuming job, game, and avocation, as it is for so many professional politicians. But it does mean politics (citizenship) as a way of living: a fact of one's life, an expected element of it, a prominent and natural role in the same manner as that of parent or worker. Strong democracy is the "politics of amateurs":

> Active citizens govern themselves directly here, not necessarily at every level and in every instance, but frequently enough and in particular when basic policies are being decided and when significant power is being deployed. Self-government is carried on through institutions designed to facilitate ongoing civic participation in agenda-setting, deliberation, legislation, and policy implementation (in the form of "common work"). Strong democracy does not place endless faith in the capacity of individuals to govern themselves, but it affirms with Machiavelli that the multitude will on the whole be as wise or wiser than princes and with Theodore Roosevelt that "the majority of the plain people will day in and day out make fewer mistakes in governing themselves than any smaller body of men will make in trying to govern them."[53]

The key process, and the center of strong democracy, is talk: the ongoing deliberation of issues that clarifies the issues themselves and the values that the community brings to bear on them. The continuous, endless process of talking it over, making decisions, and acting to solve community problems exercises and strengthens commonality. It seeks to transform the inevitable conflicts into occasions of cooperation based on that commonality. Participation and community reinforce and nurture each other as they teach people how to be citizens. To work with one's neighbors on solving a neighborhood problem, or a problem that affects all neighborhoods, is to come to know those neighbors as more than the acquaintances in the house next door. They become multidimensional beings, as family members or co-workers are, but in a way that is particular to the context, the tasks shared, and the roles played in tackling community problems. This process helps forge the *we* of community, whether the encounters are cordial or not. Something like this process can be seen in news accounts of the workings of the U.S. Senate or House of Representatives, which often reveal details of the interaction among members that suggest they see themselves as a community, if an elite one. They bond through dining, exercising, playing, and even worshipping together, even when their deliberations are markedly partisan. The sense of "we" evidently survives the friction and may even be strengthened by it.

Contrast this conception of political talk to those ludicrous "town meetings" some presidential candidates are so fond of, or the speechifying we usually associate with politicians. Perhaps no image symbolizes the impoverishment of that form of communication better than the televised scenes of politicians weighing in on matters of grave national importance in speeches supposedly delivered to their colleagues gathered in the House or the Senate chambers—the essence of what we think of as democratic deliberation. The cameras are controlled by Congress, though, and fail to show that the chambers are often empty. The videos are mainly meant for domestic consumption, fed by satellite to TV stations at home for the eleven o'clock news. This is not deliberation but public relations.

Talk in strong democracy is not one-way but entails listening as well, which is one of the most difficult and least cultivated arts of civic life. It has often been said that our culture is adversarial; we joust with words rather than lances, but still we joust. American courts and political organizations are characterized by talk aimed mainly at winning, not discerning the truth, exploring issues, or establishing bonds. Our talk emphasizes our differences and rehearses our fragmentation. In strong democracy, talk and listening are inseparable parts of the same act, and they serve to explore mutuality and build community. Listening in public deliberations may be even more difficult than listening in a marriage, but as in a marriage, it responds to effort and the perception of the same effort by others and thus increases the sense of bondedness and mutuality. That practice is part of the education into citizenship.

Strong democratic talk also prizes the emotional as well as rational. From the strong democratic viewpoint, speech that is merely meant to convey information or persuade listeners to adopt a particular stance is insufficient. Under the current system, talk aims or pretends to be strictly rational, which is congruent with the philosophical assumptions upon which its politics and economics are based—that of perfectly rational individuals self-interestedly seeking to maximize personal utility. But if those assumptions are as crumbly as we attempted to show in our critique of *Homo economicus,* then the emotional function of strong democratic talk emerges as an advantage. Besides information and rhetoric, strong democratic talk reveals to the community what its members are feeling and thus reinforces the message of their common humanity. In deliberation, the emotional content of speech conveys a measure of members' investment in the community and its actions. It puts flesh on the hard bones of rational discussion, helps illuminate the community's values, and opens a window onto its state of well-being.

More formally, Barber lists nine functions of strong democratic talk. First, it is used for *communicating interests and bargaining.* In any political system, people talk to tell each other how they feel, what they want, and what they are willing to give for it. Second, it is the medium of *persuasion,* the art of getting others to see that your perception of reality (truth) is accurate. Barber argues that these two functions of talk are recognized by all forms of democratic governance. Both are compatible with representation, and liberal democracy asks little more of democratic talk.

Strong democracy, on the other hand, sees talk as the medium for other functions as well. Third is *agenda setting.* Deciding what to talk about, and deciding who gets to decide and how, is as important in strong democracy as the talking itself. An issue that cannot be placed on the agenda is invisible, and control over the agenda is virtually control over governance. Fourth, democratic talk allows for *exploring mutuality,* which is "an informal dialectic in which talk is used not to chart distinctions in the typical analytic fashion but to explore and create commonalities."[54] It is casual conversation aimed at learning about one another and, by reciprocal exploration, creating a shared experience. In exploring mutuality, the emphasis is on relating (though not necessarily liking), not on testing the other's arguments or scoring debating points.

Fifth, strong democratic talk is a means of developing and expressing *affiliation and affection.* This is speech that is not about rationalism (debate, bargaining, etc.) but the acknowledgment and expression of connections that underlie relationships and help set bounds around the conflicts that inevitably arise among us. The language, often ritualized, reminds us that there is a fundamental "we" despite our disagreements. It is also instrumental in cultivating empathy, which is one of the features that distinguish a civic community from the legally defined relationships that characterize, say, parties to a contract.

Sixth, strong democratic talk not only helps create bonds between the members of the civic community, it also makes those bonds meaningful by *maintaining the autonomy* of the individual members. In the process of continually deliberating, discussing, and relating to each other, members find their thoughts and opinions mingling and colliding with those of others, leading to a reexamination of convictions and either their revision or their confirmation, which Barber calls "repossession." This process, invaluable in seeking to perceive truth more clearly, helps members know what they think and why they think it. It parallels adaptive management in that the perpetual testing of notions allows individuals and the community they comprise to react to changing circumstances and avoid calcification of views and values.

Seventh, the maintenance of a strong civic community requires that dissent be given enough prominence to remind the community that its occurrence warns of the danger of fragmentation. Talk serves this purpose by allowing for *witness and self-expression.* Where simple majoritarianism presumes that losing minorities will simply stay on board and maintain their allegiance to the whole community despite their loss, strong democracy institutionalizes and honors dissent. Witness and the expression of opposing opinions remind people that a free community must take diversity as a given because complete and final unity is impossible among autonomous citizens whose views are expected to evolve. Everyone will be in the minority from time to time. Honoring dissent through witness and self-expression helps avoid the alienation of the minority from the larger community, the creation of ingroups and outgroups, and the destructive efforts to avenge losses. The experience of being in the minority without being estranged from the community is one of the glues that bind the community and stabilize it.

The eighth function of talk is related to several of the earlier functions:

to allow *reconceptualization and reformulation* by the community—of agendas, of values and beliefs, of what democracy means, of what it means to be a community, of various aspects of truth itself. Particularly interesting is its value in reenvisioning the future, an act whose ongoing repetition is indispensable to the process of moving toward sustainability. It is important to see this function as one possessed and exercised by all citizens, not just leaders, and it is vital to hold in mind the power of visions. Especially in a media-driven world, to fail to talk about what future appears in store and whether the community ought to do something about it—to fail to act in the spirit of the neighbors conversing over the backyard fence about what a few degrees of climate warming might mean for their grandchildren—means surrendering to a future designed and controlled by others and shaped by forces that have only their own parochial interests at heart.

Finally, talk has one all-important function that is the end result of the first eight: *to build a civic political community* by continually educating people into citizenship, by training them formally and through participation to think publicly and to develop political judgment about what the common good ought to be.

In broadest outline, these points are the essence of Barber's theory of strong democracy. The key question is, Can theory be made reality? Can strong democracy work in the modern world?

Many theorists say no. They argue that strong democracy does not have anything like a base of modern practice. They say that people are uninterested and/or incapable of self-government and that strong democracy is impossible in a world of eroding nation-states, ascendant transnational corporations, and a globalizing economy, because it is suited only to isolated, stable, small-scale communities whose time has gone. To assess these objections, it is necessary to decide on how high to raise the bar. Certainly it is ridiculous, for example, to contemplate a citizen assembly for the whole country. However, it is not ridiculous to envision citizen assemblies at the neighborhood level, which could be networked into a community of communities and thus a system that could inject citizen input directly into the policymaking process at the national level. The Raetian Republic, which thrived in Switzerland for nearly three hundred years in an era when the fastest form of communication was still a man on a horse, offers an example for such a system. In other places and times, various other mechanisms embodying one or more features of strong democratic governance in real-world systems have been tested and found to be workable and useful. Of course, some of these efforts could be called qualified successes or even failures (see chapter 6). We conclude this chapter with an outline of Barber's suggestions for implementing strong democracy.

How Strong Democracy Might Actually Work

Imagine the Las Vegas man with the illegal water sprinkler, whose story was recounted earlier, living in a community organized on the principles of strong democracy. Chances are that he, or his wife or brother or neighbor, might occasionally attend meetings of the neighborhood assembly. Water topics no doubt would frequently appear on the agenda; in fact, since the lack of water is a defining characteristic of the Southwest, it probably would be an ongoing item of concern and discussion. Particular matters relating to water availability and distribution would be debated periodically during regional interactive television town meetings, and would come up for resolution by means of local or regional referenda. The man might even have been chosen by lottery to sit on the local water resources board. He would know more about the watershed he lived in, where his water came from, and its true monetary and ecological costs. He would inevitably have a fuller understanding of these and other factors (irrigation, for example) that shape the precarious dependency relationship his city and region have with water. By the same token, so would everybody else on the block. The rules about water usage, rather than seeming to be imposed by a remote and arbitrary bureaucracy, would be part of a community covenant, informed by ecological understanding, that he had taken an active part in drawing up. His nonsensical outrage ("You people are trying to turn this place into a desert") would be hard pressed to survive his education into citizenship in an arid community. He might be less inclined to abuse his water rights, and the neighborhood might be more inclined to gently point out the error of his ways.

This illustrates how strong democratic structures could work to build communities of citizens whose active political participation in self-governance was informed by confrontation with ecological reality. Strong democracy argues that community is fostered by participation, and participation by community. What is necessary are institutions to promote both, institutions that are workable, compatible with existing structures of representative democracy (which should not be destroyed, only modified), and capable of addressing the various obstacles to participation.

Barber proposes a set of institutions to support the talk, decision-making, and action at the heart of strong democracy. His proposals are prescriptive and unitary; take them all or forget it, he says, because they reinforce each other and would not survive if adopted piecemeal. This is not necessarily true, and is somewhat at odds with his advice to layer strong democratic structures on top of the existing order rather than sweep it away. In any case, the likelihood of persuading any community to implement his program in toto seems remote, while the success of some communities in experimenting with various partial strong democratic structures and ideas suggests that wholesale adoption may be unnecessary.

TALK INSTITUTIONS. Talk institutions would include, first, *neighborhood assemblies,* probably organized into regional associations. The assemblies would vary in physical size, depending on whether the district was a city, suburb, or rural area, but would be large enough to encompass several thousand citizens.[55] They would serve at first to ensure accountability of local officials, as an institutional ombudsman for the hearing of grievances and disputes, and as arenas in which to discuss local issues as well as referenda affecting the region and the nation. Barber advocates limiting the assemblies to these functions for an unspecified period of time to develop civic competence and confidence before allowing them to assume powers of local self-rule (although, if strong democracy depends on people's decisions making a difference, this arrangement is unlikely to inspire much enthusiasm[56]). Meetings might be weekly and scheduled at such times as to encourage workers and parents to attend. Agendas would be set by the participants. The assemblies could meet in local schools or recreation halls. Eventually, they would become small-scale legislative bodies with power to regulate local affairs and deliberate on regional and national matters by means of referenda.

This regional and national activity would be mediated by means of *television town meetings* and a *civic communications cooperative*. Modern electronic communications media offer great potential for wide-scale communication and deliberation, though at some cost in intimacy. Certainly televised deliberations would involve only selected individuals, though they could be broadcast to essentially everyone. Television has already been used successfully to promote meaningful political debate in New England, California, Hawaii, Maryland, New Zealand, and many other places. With appropriate safeguards, various electronic media could also be used for debate-centered polling and voting. To avoid abuses and ensure that the process is not left to the market, Barber argues for a national civic communications cooperative; it would be set up as a publicly controlled body responsible for experimenting with new forms of civic broadcasting, issuing guidelines for the various forms of public talk, overseeing electronic polling and voting, and in other ways ensuring the integrity of electronic civic communication. It would be independent of private broadcasting and have no authority over it.

In the interests of ensuring equal access to useful information and promoting civic education, the communications cooperative would also operate an *interactive electronic forum*. Barber proposed a civic videotext service when his book was originally published, but technology has overtaken his suggestion, and these days dedicated World Wide Web sites, news groups, and chatrooms are more appropriate. They would provide widely accessible civic information in the form of news, testimony, discussions, and data of current issues facing the community, the region, and the nation. Likewise, a *civic education postal act* would subsidize the mail distribution of printed materials relevant to civic affairs and civic education.

Finally, although direct democracy and universal participation are the goals, some compromise may have to be made in certain circumstances, as in the case of regional assemblies or when neighborhoods have too many members for everyone to take part. Representatives might be chosen in those cases, but to ensure that everyone eventually has a role, selection could be by lottery and representatives would rotate in and out of the assembly. Many nontechnical local offices often filled by election—boards of education and zoning and various commissions, for example—could also be filled by lot and held on a rotating basis.

DECISION-MAKING INSTITUTIONS. First among the decision-making institutions are the *referenda* mentioned above. Many states already employ statewide initiatives and referenda to make policy. These are not used at the national level in the United States—a legacy of the Madisonian terror of popular mismanagement. Barber's proposal defends against those fears by incorporating a carefully structured initiative/referendum process that requires at least two readings (votes) by the citizenry (to guard against impetuosity), related civic education activities in neighborhood assemblies and through the media, and a format that allows for several responses to referendum questions in addition to yes and no. The additional responses allow people to express qualified support or opposition, or to request further debate and deliberation.

Interactive television might allow for *electronic balloting* as a means of promoting thoughtful deliberation and "empathetic forms of reasoning" during neighborhood assembly debates or regional and national referenda. Electronic balloting, however, might become a rapid polling mechanism rather than a debate-steering device, a danger Barber acknowledges without spelling out how it could be avoided. A carefully modulated debate structure designed to lead up to a referendum, with intermittent polling throughout as a means of gauging the will and responses of the people, might avoid the dangers of impulsive voting.

As mentioned above, *selection by lot* (sortition) of delegates to regional assemblies or of certain officeholders would also further democratize civic decision making. This tradition dates back at least to the ancient Greek *polis* and survives in the selection of jurors. When applied to assemblies, random selection of citizen delegates could help address the problem of scale and the tendency for wealth to dominate the electoral process. Especially when coupled with periodic rotation of officeholders, it would also help keep the power-mad out of office and thus institutionalize the wisdom of those hunter-gatherer groups who deliberately deny power to those who seek it most avidly. Public offices not requiring much special expertise could be filled by lot with citizens given some basic grounding in the responsibilities involved. These could include boards of assessors, planning, zoning, and licensing; county commissions; and voter registries. Officeholders would receive per diem compensation.

Vouchers have been widely debated in the United States as a means of improving school performance by increasing competition. Despite reservations, Barber supports experimenting with vouchers to promote democratization. When public school systems are failing, for example, a voucher system could motivate the schools and their attendant bureaucracies to seek excellence or risk going out of business while simultaneously engaging parents in their children's education by requiring them to compare school programs and choose the one best suited to their needs—one important avenue to broader civic engagement. Such plans would have to be coupled with measures to prevent the rich from supplementing vouchers with their own wealth and to subsidize transportation so that the poor would not be geographically constrained. The dangers, however, are significant. Privatizing education means subverting its function as a means to promote the public good because people are thinking only in terms of their own children's welfare, not the welfare of all the community's children. Vouchers also bypass neighborhoods and thus undermine the neighborhood as the citizen's civic base. This mix of virtues and vices argues for cautious experimentation with vouchers.

ACTION INSTITUTIONS. Because, in strong democracy, politics is not merely willing and deliberating but acting as well, institutions for focusing hands-on action that are compatible with strong democratic thought are necessary. Chief among these would be the establishment of a national service requirement of a year or two, either in the military or a civilian venue, as is already done in Germany and Switzerland. This would help redress the balance between democratic rights and democratic obligations in civil society, as well as demercenize the military and make it less of a last-resort source of opportunity for the poor. Universal service would also socialize citizens of all races and classes together in a context of civic cooperation and education. Finally, it would help balance the tendency of many of the other structures of strong democracy toward a narrow, local focus by engaging everyone in work of national significance. Options for those entering national service, besides the military, might include an urban project corps (park restoration, graffiti cleanup, traffic control, daycare, elderly assistance); a rural project corps (conservation and reforestation, flood control, road maintenance, fighting forest fires), an international corps much like the Peace Corps and with similar functions, and a special services corps to provide the other corps with people skilled in medicine, communication, construction, administration, and so on.

Barber also suggests experimenting with various forms of workplace democracy and investigating ways that architecture and city planning could be employed to promote a sense of neighborhood without creating enclaves and attempt to restore the socially interactive character possessed by many traditional neighborhoods.

Finally, even though strong democracy is a reaction to the checks and balances and the institutional separation of citizens found in the procedural republic, it, too, requires built-in constraints. The strong democrat must find ways to "institutionalize regret: to build into his reforms limits on the will to change and to build into mechanisms of public choice limits on all political will."[57] For example, the referendum process needs to be preserved from rashness by requiring full debate via the civic communications network, two readings for all propositions, and the possibility of a congressional veto.

———

Barber's is not the only strong democratic vision. There are others that differ radically in many particulars.[58] Murray Bookchin, for example, argues for a system he calls confederated municipalism that is modeled in many ways on the city-states of pre-Renaissance and Renaissance Europe. Far from seeking to link neighborhoods to the nation in a hierarchy of aggregated political structures, Bookchin would just as soon destroy the modern state; any power city-states acquire will come at the state's expense anyway, he says. He is vehemently opposed to any form of electronic democracy and even to referenda, preferring reliance on the face-to-face assembly. Connected by computer or interactive television to an assembly debate but still stuck at home, Bookchin argues, the individual is "left to his or her own private destiny in the name of 'autonomy' and 'independence'" and thus "becomes an isolated being whose very freedom is denuded of the living social and political matrix from which his or her individuality acquires its flesh and blood." It is "interdependence within an institutionally rich and rounded community—which no electronic media can produce—that fleshes out the individual with the rationality, solidarity, sense of justice, and ultimately the reality of freedom that makes for a creative and concerned citizen."[59]

Strong democracy is not an off-the-shelf proposition, despite the tidy packaging of the various proposals, no matter what form it might take. The emphasis on contingency and process that suit it so well to the demands of sustainability also mean that its structure cannot be precisely defined for every community or for all time. It must evolve with experience; it must be reinvented from moment to moment. Barber himself emphasizes layering the mechanisms of strong democracy over the existing structure of liberal democracy, not replacing the one with the other as in a coup d'état. Some of those mechanisms will work, some won't. Communities would be wise to apply the principles of adaptive management to their government as well as to their economies and social lifestyles.

Strong democracy is also not without risks. There is the very plausible danger, were reforms to be implemented, of indifference or disillusionment and lack of staying power in a populace unaccustomed to governing themselves and among whom even a simple and limited civic responsibility like jury duty is often considered a burden. Education into citizenship might have to be aggressively promoted before it reached a self-sustaining level. On the other hand, if reforms succeeded in involving more people in governance, the result might be sluggish government. Fervent citizen involvement can mire any proposal in lengthy debate over minutiae. There are also examples, such as the sectional assemblies of the French Revolution, in which strong democratic governance took on an unnervingly brash and headstrong style.

Whether this or any form of strong democracy would work better than the system now in place, or work at all, is a matter of speculation. We have argued that some form of strong democratic governance would be more conducive to sustainability because it would bring more people face-to-face with the problems posed by unsustainable practices as well as give them the power and responsibility for acting to address those problems. We believe, too, that it is a truer form of democracy than the liberal democracies that are the norm in the West. For these reasons it is worth trying—or trying again and more widely, since many examples of successful direct democracies can be drawn both from history and current practice. Chapter 6 examines some of those efforts.

Notes

1. D. Rothman and S. de Bruyn, "Probing the environmental Kuznets curve hypothesis," *Ecological Economics* 25 (May 1998): 143–145.

2. R. Putnam, *Making Democracy Work: Civic Traditions in Modern Italy* (Princeton: Princeton University Press, 1993), 173, 175.

3. K. Lee, *Compass and Gyroscope: Integrating Science and Politics for the Environment* (Washington, DC: Island Press, 1993), 10.

4. R. Samuelson, "Why Clinton hangs on," *The Washington Post,* April 1, 1998, p. A19. Ironically, that well-being might actually be enhanced by a system structured to encourage more political participation. A study by two economists at the University of Zurich suggests that people with abudant opportunities to take part in direct democratic activities are happier than those with fewer such opportunities (B. Frey and A. Stutzer, "Happiness and institutions" draft paper, April 16, 1999).

5. Democratic systems are not necessary for economic success, though, as such autocratic but prosperous market systems as Singapore demonstrate.

6. J. Krosnick, P. Visser, and A. Holbrook, "American opinion on global warming: The impact of the fall 1997 debate," *Resources* 133 (fall 1998): 5–9.

7. Carbon dioxide is the chief greenhouse gas after naturally occurring water vapor.

8. Percentages calculated from data in U.S. Energy Information Administration, *Emissions of Greenhouse Gases in the United States, 1997,* DOE/EIA-0573(97), October 1998, table ES3.

9. U.S. Energy Information Administration, *International Energy Annual, 1996,* DOE/EIA-0219(96), February 1998, table H1.

10. World Resources Institute et al., *World Resources, 1998–1999: A Guide to the Global Environment* (New York: Oxford University Press, 1998), table 16.1.

11. R. Sanchez, "West wages a new sort of turf battle," *The Washington Post,* May 16, 1999, p. A3.

12. S. Funtowicz and J. Ravetz, "A new scientific methodology for global environmental issues," in R. Costanza, ed., *Ecological Economics: The Science and Management of Sustainability* (New York: Columbia University Press, 1991), 137–152.

13. Ibid., 138.

14. M. Waldrop, *Complexity: The Emerging Science at the Edge of Order and Chaos* (New York: Simon and Schuster, 1992), 82.

15. M. O'Connor, S. Faucheux, G. Froger, S. Funtowicz, and G. Munda, "Emergent complexity and procedural rationality: Post-normal science for sustainability," in Costanza et al., *Getting Down to Earth*, 229.

16. P. Montague, "The waning days of risk assessment," *Rachel's Environment and Health Weekly* 652, May 27, 1999.

17. Ibid., 146.

18. O'Connor et al., "Emergent complexity and procedural rationality," 226.

19. Ibid., 149, 151.

20. Ibid., 237.

21. Ibid., 149.

22. Ibid., 235.

23. Lee, *Compass and Gyroscope*, 8.

24. Ibid., 91, 92.

25. Ibid., 92.

26. Ibid.

27. Ibid., 100.

28. L. Hempel, *Environmental Governance: The Global Challenge* (Washington, DC: Island Press, 1996), 234.

29. Ibid., 6.

30. Republicanism in this context is not to be confused with the earlier distinction made between republican and democratic forms in terms of representational versus direct government. Here, Jeffersonian republicanism embodies the direct democratic form, while the federalism of Madison and Hamilton prefigures the centralized, representational form we have today.

31. D. Kemmis, *Community and the Politics of Place* (Norman, OK: University of Oklahoma Press, 1990).

32. Ibid., 11, 12.

33. Ibid., 15. The Civil War might be taken as evidence that political conflict could not be contained either by the checks and balances built into the Constitution or by the frontier. In any case, the frontier was declared closed in 1890, with the admission of Idaho and Wyoming to the Union. Western migration and settlement continued, but symbolically at least, the geo-

graphic safety valve was removed. Kemmis asserts that it was replaced by two other safety valves: America's expansionist drive to empire, which superficially united people in their common patriotism, and the creation of the modern regulatory bureaucracy, which prescribed people's relations with each other through rules and regulations. We might add to this list modern consumerism and the mantra of endless growth, which help pacify the discontented with the promise of a piece of the ever-enlarging pie.

34. Kemmis, *Community and the Politics of Place*, 30.

35. Cited in Kemmis, *Community and the Politics of Place*, 31.

36. See B. Barber, *Strong Democracy: Participatory Politics for a New Age* (Berkeley: University of California Press, 1984).

37. Ibid., 132.

38. Ibid., 8.

39. See M. Berry, K. Portney, and K. Thomson, *The Rebirth of Urban Democracy* (Washington, DC: Brookings Institution, 1993), chapter 9.

40. Barber, *Strong Democracy*, 20.

41. Ibid., 21.

42. Ibid., 90.

43. Cited in Barber, *Strong Democracy*, 90.

44. Barber, *Strong Democracy*, 120, 121.

45. Ibid., 129.

46. Attributed.

47. Barber, *Strong Democracy*, 126.

48. Ibid., 135.

49. Ibid.

50. Ibid., 136.

51. Ibid., 145, 146.

52. Ibid., 147, 148, 152.

53. Ibid., 151.

54. Ibid., 183.

55. Recall that Plato thought 5,000 families was about right. Barber says 5,000 to 25,000 people. Other writers offer other numbers, and there seems to be no consensus on the matter. It's likely that preferences will differ from place to

place and that the ideal size, if there is one, would be identifiable only through experience. One constraint would be the sheer practicality of having more than a few hundred people gather in a hall at one time and still be able to participate meaningfully. This is not inconsistent with the ideas of strong democracy, since there is no expectation that every person would attend every meeting and speak every time.

56. Barber shares this reservation: "When participation is neutered by being separated from power, then civic action will be only a game and its rewards will seem childish to women and men of the world" (*Strong Democracy,* 236).

57. Ibid., 308.

58. Thad Williamson has compiled and synopsized several dozen schemes for overhauling current economic and political structures; many of them include strong democratic features. See *What Comes Next? Proposals for a Different Society* (Washington, DC: National Center for Economic and Security Alternatives, 1997).

59. Bookchin, *From Urbanization to Cities: Toward a New Politics of Citizenship* (London: Cassell, 1995), 225, 226.

Chapter 6

The Once and Future Democracy

This chapter aims to render strong democracy plausible. It seeks to show by example that various ideas and pragmatic features central to strong democratic governance have been proven viable by history and/or contemporary practice. We ask three questions: Has direct democracy ever worked? Can it work in the modern world? Can it make for better governance, and thus, presumably, more effective solutions to the problems of sustainability? In all three cases, our examples suggest that the answer, with some caveats, is yes.

Direct democracy has many incarnations and offers several examples, both famous and obscure, from the past and the present. While none of the examples discussed below reveals a form of strong democracy embodying all the features described in chapter 5, individually they show many of the elements of direct democratic practice and thus illustrate the practicality of their implementation in the real world. They also remind us that, as Churchill famously observed, democracy in general is "the worst form of government."[1] Direct democracy certainly has its share of defects, both historical and contemporary. The glories of the fabled Athenian *polis* rested on a foundation of slavery. New England town meetings of the colonial era, sentimentalized in popular memory, fell short of the modern ideal by sharply restricting the right to vote and by sometimes serving as machinery for maintaining the status quo rather than hosting the creative clash of ideas and interests.[2] Other examples, however, show just as plainly that an engaged citizenry is strongly correlated with the effectiveness and responsiveness in government that is a prerequisite to addressing sustainability problems.

Historical Direct Democracies

Let's begin with some snapshots of direct democratic societies that flourished in the past:

PERICLEAN ATHENS. The prototypical democracy is the fifth-century B.C. *polis* of Athens, which reached its greatest heights, ironically, under the nominal dictatorship of Pericles.[3] Though there was a superstructure of government, the basic governing body was the assembly *(ekklesia)*, which consisted of every eligible voter in Attica, the peninsular province surrounding Athens. Eligibility in Periclean Athens meant being at least twenty-one years of age and born of two free Athenians, and not being a woman, slave, workingman, or resident alien. Thus, 43,000 people could vote out of an estimated population of 315,000. Though admittedly limited, this franchise gave serious rights and duties to those who held it. Citizens were formally equal to one another regardless of class or occupation. All were liable for service as judges or magistrates, which were selected by lottery or by rotation.

The body politic was split between the oligarchs and the democrats, which roughly mirror the traditional divisions of modern American politicians into Republicans and Democrats. The former in each case tend (or tended) to be allied with moneyed interests, while the latter found support among the lower classes and often advocated policies designed to redistribute wealth. The assembly met about four times a month. Only two thousand or so of the eligible voters usually attended, owing to the difficulty of transportation, and because many of the oligarchists lived in the hinterlands, the mainly democratic residents of Athens thus came to dominate the assembly. The meetings, which began at dawn with the sacrifice of a pig, took place under the open sky. Originally, the assembly met at the *agora,* which eventually became the site of shops and temples. By the time of Pericles, the gathering had moved to a nearby hillside, but the *agora* remained the social center of the city and a place of perpetual discussion and discourse.

Legislation was introduced in the assembly, but before consideration there it went to the *boule,* or council. In Pericles' time, this was a Council of Five Hundred, selected by lot and rotation. Once a citizen had served his year's term, he was not allowed to serve again until all citizens had served their terms. Besides reviewing and reporting on proposed bills, the council supervised city administrators, minded the budget, and established foreign policy. The assembly could accept or reject the council's legislative recommendations and review and override its executive decisions.

Once the council had reported out a bill, it went to the assembly for debate. Any citizen could speak, but the assembly was a tough crowd: "Only trained orators avail themselves of the right to speak. . . . [The assembly] laughs at mispronunciations, protests aloud at digressions, expresses its approval with shouts, whistling, and clapping of hands, and, if it strongly disapproves, makes such a din that the speaker is compelled to leave the . . . rostrum." Members wishing to introduce legislation were wise not to do so frivolously, for they would be responsible for the results: "If these are seriously evil another member may within a year of the vote invoke upon him the . . . writ of illegality, and have him fined, disenfranchised, or put to death."[4]

The citizen who populated these institutions was forged by the process the Greeks called *paideia,* the education into citizenship that was, in Murray Bookchin's words, "a deeply formative and life-long process whose end result made [the citizen] an asset to the *polis,* to his friends and family, and induced him to live up to the community's highest ethical ideals." *Paideia* "expresses a creative integration of the individual into his environment, a balance that demands a critical mind with a wide-ranging sense of duty."[5] This rounded striving for excellence in a civic context led to an extraordinarily deep concept of citizenship:

> [Citizenship was] an ethos, a creative art, indeed, a civic cult rather than a demanding body of duties and a palliative body of rights. At his best, the Athenian citizen tried not only to participate as fully as possible in a far-reaching network of institutions that elicited his presence as an active being; the democracy turned his participation into a drama that found visible and emotional expression in rituals, games, artwork, a civic religion—in short, a collective sense of feeling and solidarity that underpinned a collective sense of responsibility and duty.[6]

How different from the modern notion of the good citizen: vote, pay your taxes, mind your own business. And many don't even vote. This is not to say that Athenians were paragons of virtue—Will Durant comments that "the Greek might admit honesty is the best policy, but he tries everything else first"[7]—only that their democratic system superbly trained its citizens to a role they not only took seriously but lived fully and unselfconsciously. Pericles died in 429, but the Athenian democratic system survived for some time, intermittently subverted and restored, then increasingly compromised by a string of demagogic rulers. Athenian independence lasted until 338 B.C., when the Macedonians under Philip, father of Alexander the Great, routed the Athenian army at Chaeronea.

THE ITALIAN MEDIEVAL COMMUNES. In the twelfth and thirteenth centuries, Italy responded to the violence, anarchy, and political turmoil of the Middle Ages by gradually giving birth to two strikingly different systems of government. In the south, transplanted Norman mercenaries established an autocratic state centered in Sicily that, for a time, was noted for its enlightened, efficient, and culturally fecund rule. Though its founding rulers Roger II and later Frederick II retained nearly absolute power, they built a state on a system of codified administrative law more sophisticated than any since Roman times, practiced religious toleration, and got rich on state trade monopolies. Their power and that of their successors declined in the following centuries, devolving to the nobility. The kingdom slipped into the feudal patterns common elsewhere in Europe, marked by a hierarchical social structure with a mass of desperately poor and powerless peasants at the bottom and a wealthy aristocracy on top.[8]

The towns to the north, in contrast, went off in a radically different direction, toward self-government in the form of communal republicanism. For self-protection and mutual assistance, groups of city dwellers formed voluntary associations in which they pledged to cooperate economically and in defense. By the twelfth century, communes had evolved from these associations in most of the large northern and central Italian towns, including Florence, Venice, Bologna, Genoa, and Milan. As in Athens, suffrage was by no means universal. The features of government varied from place to place; often it was republican in form but at other times clearly democratic, with a popular assembly consisting of citizens mutually sworn to the protection of their common interests. The oath "committed the citizens of the commune to orderly and broadly consensual ways of governing themselves with a decent respect for individual liberty and a pledge to their mutual defense."[9] In any case, all the towns were governed more democratically and on the basis of a wider distribution of power than anywhere else in Europe at the time. A city council, for example, might have thousands of members drawn from all ranks of society, including workingmen, not just the traditional elite. Later, craftsmen and tradesmen also formed associations (guilds) and agitated for even broader distribution of power. These groups were part of a larger movement toward greater equality:

> Beyond the guilds, local organizations, such as *vicinanze* (neighborhood associations), the *populus* (parish organizations that administered the goods of the local church and elected its priest), confraternities (religious societies for mutual assistance), politico-religious parties bound together by solemn oath-takings, and *consorterie* ("tower societies") formed to provide mutual security, were dominant in local affairs.... This rich network of associational life and the new mores of the republics gave the medieval Italian commune a unique character precisely analogous to what . . . we termed a "civic community."[10]

In parallel with these developments came an increased reliance on written agreements, contracts, covenants, negotiation, mediation, and the law in general, with the combined effect of creating an ordering principle that was a powerful alternative to the sword wielded by a baron or a king: "At a time when force and family were the only solutions to dilemmas of collective action elsewhere in Europe, citizens of the Italian city-states had devised a new way of organizing collective life."[11] And the payoff lay not just in civic pride but in commercial expansion as well: the improvement in civic order encouraged merchants and traders to look beyond local markets and seek wider ones, first within Italy and then outside its borders. In short, they founded capitalism—thus demonstrating that social capital is the sine qua non of economic capitalism. Capitalistic expansion also depended on credit, which the Italian republics invented and which could not have emerged without the trust and confidence built up through the institutional relationships and networks of civic republicanism.

During the fourteenth century, Italy was wracked by powerful disruptive forces: factionalism, the rise of petty tyrants backed by mercenary armies, the plague. Disease alone killed more than one-third of the populace in a single year (1348), leaders and citizens alike. Repeated epidemics followed in subsequent years. So devastating were these upheavals that a number of the communal republics in the far north succumbed to despotism, more than two centuries after their founding, and the rest ceased to exist as communal republics as the following centuries saw Italy become a battleground where Europe's great powers fought for continental supremacy. The communes incrementally surrendered their power to executive officials over time:

> Popular assemblies began to vote increased emergency powers to their *podesta* and captain; the same men were repeatedly confirmed in executive office, and their tenure was extended from one year to five, from five to ten, and ultimately to lifetime office. Soon the leader's son supplanted his father after death and dynasties began to turn elective offices into mere charades.[12]

Still, the communal republics left a remarkable legacy. The Italian city-states served as a civic laboratory for the rest of Europe, and the experiments begun in them were later repeated and embellished north of the Alps.[13] The republics also made an imprint on Italian civic sensibilities that seems to have lasted to this day. At the beginning of the fourteenth century, according to political scientist Robert Putnam, the regimes in Italy could be described roughly as four types, running from south to north (see fig. 6.1): the Norman feudal monarchy; the papal states, consisting of a mix of feudalism, despots, and republican traditions; the surviving republican communes; and in the far north the once-republican cities that had lost their self-governing structures (but not their memories and sentiments of self-governance) to tyrants. In effect, the spectrum runs roughly from minimal civic engagement to strong civic engagement as it runs from south to north. Putnam argues that, to a very considerable

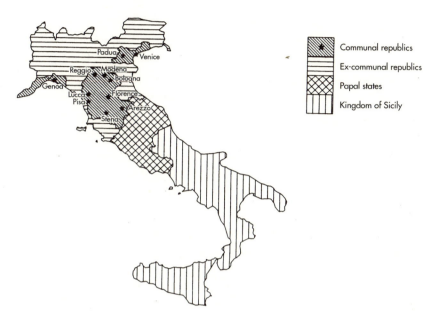

Figure 6.1. Republican and Autocratic Traditions: Italy, c. 1300

extent, the pattern can still be seen in modern Italy: the south is marked by hierarchical, patron-client–type relationships (and not only in the Mafia) and relatively ineffective government, while in the north, where the ancient traditions of republicanism, lateral networks, and engagement persist, government is much more effective and efficient. Though a sense of true citizenship is time-consuming and difficult to cultivate, it is a hardy flower. The effect of this legacy on modern Italy is touched on in the following section on direct democracy in modern times.

THE REPUBLIC OF RAETIA. The Italian city-states of the high Middle Ages are well known compared to the Raetian Republic. Yet it was a remarkably successful experiment in direct democracy that lasted nearly three hundred years, from about 1524 until Napoleon forced its unification with Switzerland in 1799. Its democratic roots go back at least to the seventh century.[14]

The Republic of Raetia is now the southeastern Swiss canton of Graubünden. It is a place of diversity, both in its landforms and its people. The mountainous landscape isolates it from the rest of Europe, yet its central location and possession of half the major passes in Switzerland have made it the focal point of much European political conflict. The 150 separate valleys, each with its own topography and features, were inhabited by a people divided among Germanic, Raeto-Romansch, and Italian groups, and about evenly split between the Roman Catholic and Protestant faiths. Most of the people lived in high valleys; even as recently as 1970, less than 3 percent of the Swiss population lived above 1,000 meters elevation, but half the Raetians did. Poor soils and a lack of natural resources forced the inhabitants into a largely pastoral (herding cattle) way of life, which hardened them and, during the six-month winters, afforded ample time to practice fighting skills and politics. Consequently, Raetians became known both for their ferocity as mercenary soldiers and for their vigorous communal politics.

That politics was marked by interdependence and mutualism. The harsh environment drove the Raetians into each other's arms, so to speak: "Only collective labor and common decision making afforded protection against nature's hostility, the more so in earlier times before technology became available in the war to subjugate nature. Thus, as the hardness of life molded a man's sense of autonomy, it also compelled him to cooperation and collective action."[15] The result, in the early Middle Ages, was the "association of the common," a collective organization whose purpose was to exercise the right of usage (not ownership—that was the king's or the nobility's right) and administer the common lands and resources. Though also found elsewhere in Europe, the commons associations persisted and remained more important in Raetia, and in some instances resisted privatization until the 1800s. Privatization eventually reduced the effective scope of common usage to grazing rights, but these were institutionalized in a way that allowed members of the association to herd their cattle across all lands, regardless of ownership, within the association's jurisdiction. These valuable rights accrued to any member of the association, which meant any man who lived within its geographical boundaries regardless of status or rank, and led to a unique social environment:

> Free peasants, descendants of the original unfree class of Roman colonists . . . , local propertyowners, and land-holding noblemen with vassal status in the Frankish Empire

found themselves in a uniquely egalitarian situation. Bound together by their common status in the association and by their common rights of usage over all collective resources, they developed a sense of equality wholly at odds with the hierarchical, status-conscious thrust of medieval feudalism.[16]

Encroaching feudalism attacked the associations on several fronts, as landowning clergy and nobility attempted to usurp and tax the common lands and undermine the associations' autonomy. The institutional response of the free peasantry was, in general and over time, to reduce the size of the associations to village communities rather than regions, and to become politically more assertive. The securing of such crucial political rights as power over the apparatus of criminal law, till then held by the feudal lords, took centuries. But when the process was finally completed, the autonomous village communes that had evolved from the ancient common associations during this struggle formed the basis of the Raetian Republic. The republic's independence, achieved piecemeal by the individual communes, was formally consolidated when the communes signed a pact of confederation in 1524.

Raetian political organization was not simple, consisting of several overlapping levels of jurisdiction grafted on to one another by history and circumstances. At the bottom was the village neighborhood (227 in all); then came the 49 jurisdictional communes of several villages each, the 26 so-called higher jurisdictions left over from feudal times, the three leagues that formed the republic, and one executive body (the heads of the three leagues). The real power lay with the communes and the neighborhoods, and their mode of political action was face-to-face democracy. Citizens and neighbors met formally and informally, in more or less continuous fashion, to set taxes, distribute communal income, build and maintain roads and paths, administer common lands, modify the rules of citizenship, choose men for local and regional office, and generally run their own affairs. There were certain administrative posts, but these were often chosen by lottery, and no important powers were given to officeholders.

Political life did not end with talking and voting; it meant common work as well:

> Active participation in the communal assembly was only the
> beginning of the collective work on which the life of the
> neighborhood depended. For example, the decision to build
> a new road could not be made in a splendid flurry of demo-
> cratic spirit and then forgotten, left to some engineer corps
> to complete. To will the road into being, as it were, entailed
> building it. Those who willed it built it, and their labor was
> regarded as an expression of commonality for which no
> compensation was required.[17]

With the widespread use of the referendum, which was probably
invented in Graubünden, the Raetians extended their democracy to the
national level. Unlike typical modern referenda, in which voters individually consider a single question and vote it up or down after wrestling (or
not) with their own consciences, Raetian referenda were decided by the
collective votes of the jurisdictional communes, which deliberated heavily
and generally sought internal consensus. The referendum was used to
make certain unusual state decisions that other countries usually left
to the executive, such as not only declaring war but also deciding where to
send the troops, with sometimes paralytic results. Domestic matters were
left mainly to the communes. Though not always successful, the referendum was still "intended to minimize cleavage, to maximize intercommu-
nal consensus, and to make possible at the level of the republic the sort of
consensus vital to the neighborhood. Commonality was the aim, not
compromise."[18]

Raetian referenda differed from modern ones in other key dimensions, such as the extent to which anybody could initiate one (even for-
eigners, before abuses ended the option), and, above all, the extent to
which an issue could be responded to in many ways, not only with a yes
or no. The communes often replied to referendum questions with lengthy
essays that probed, mused, and expressed doubts and alternatives. This
process often resulted in a muddle of incompatible responses to the questions, and to work through these was the job of the executive commission.
Not a prescription for efficiency—but then, participation, not efficiency,
was the supreme aim.

The Raetian experience shows that direct democracy is not proof against ignorance and stupidity. Raetian freemen on occasion tended toward an exasperating mule-headedness, as in the following story:

> A foreigner . . . traveling the main road, comes upon a local cowherd running his cows nonchalantly through a freshly plowed field right next to the road. The foreigner remonstrates gently, pointing out the wanton and unnecessary damage being done to the field and suggesting the use of the road. But the Bünder cowherd, with all the force of a thousand-year tradition of common grazing behind him, retorts, "And what business is it of yours? We are not in your country now! We are a free people, and we do exactly what we want to do!" The tone is disapproving, but the average Raetian reading the story would be more likely to swell with pride at his brother's fettle than to blush with shame for his mindless obstinance.[19]

Despite the inefficiency of the system—and its vulnerability, as with many systems, to fragmentation, arrogance, xenophobia, corruption, mob behavior, and other troubling tendencies—Raetia survived nearly three centuries in an era when various forces for subjugation buffeted the tiny republic repeatedly. The citizens of Raetia were as free, or freer, than any other people have ever been to shape and realize their destinies as members of autonomous communities.

NEW ENGLAND TOWN MEETINGS. The famous town meetings of New England originated in the Massachusetts Bay colony in the 1620s[20] and have lasted to this day, making them one of the longest-running forms of direct democracy. The meetings, especially those of the colonial era, are something of a benchmark in discussions of direct democratic practices. Thomas Jefferson called them "the wisest invention ever devised by the wit of man for the perfect exercise of self-government, and for its preservation."[21] That sort of accolade may be the source of their hold on the popular imagination, though they were not necessarily the exercises in pure Jeffersonianism they are sometimes thought to be.

For one thing, suffrage was restricted to adult white males. For another, after the early colonial period, the town meeting was not the only governing body. The colonies also had general assemblies, and many colonies assigned certain responsibilities to other local bodies. In Connecticut, for example, the town meeting was supplemented by separate meetings of the society (the governing body of the church parish) and of the freemen (who owned property but held no shares in the common lands). A man might be a member of all three bodies. Societies were concerned with running schools, building meetinghouses, hiring preachers, and guarding morality—matters that openly mixed politics and religion. Freemen met twice yearly, in April to elect delegates to the colonial assembly (representing the freemen, not the town per se) and in September to nominate candidates for governor and deputy governor. A third shortcoming was a tendency toward oligarchy and long-term preeminence in town affairs by dominant elite families, especially in the smaller, more rural towns.[22]

On the other hand, suffrage is never universal, power is always divided and stratified in democratic societies, and all forms of democracy must be wary of usurpation by special interests. Whatever their flaws, the meetings allowed colonial citizens to exercise considerable local autonomy. The meetings afforded all eligible voters the opportunity to exercise directly the right of governance. Meetings typically were held one to three times a year, unless a compelling matter arose requiring a special gathering. Meetings were more frequent in new towns than in mature ones, often as many as one a month. The main business of the meeting was the election of officers and the approval or rejection of their decisions; however, "Although the meeting usually chose not to involve itself in decision-making beyond electing officers, its frequent flurries of activity during controversies left no doubt where internal power ultimately lay."[23] When a town held only one main meeting in a year, that meeting usually fell in December.

Meetings perhaps made up in length what they lacked in frequency, with the main meeting often running two days or more. The tone of meetings varied according to the attendance and the degree of controversy surrounding the items on the warrant (agenda). The duration of the meetings and the surviving descriptions of votes and proceedings make it clear that much deliberation occurred in them. Consensus was the aim, and was often achieved; some towns went for years with nothing but unanimous votes before recording an issue decided by majority. Occasional conflict at certain Connecticut town meetings prompted the general assembly to ban unruly behavior and uncivil speech at meetings. Votes could be by voice, hand-raising, or secret ballot, according to need.

Perhaps obeying some universal law, the number of town officials tended to increase over time. Originally, the roster of officers included several selectmen (as executives), two constables (police), a town clerk to record deeds, marriages, births, deaths, and so on, two ratemakers for making property tax assessments, two surveyors for fence maintenance, and a few packers and sealers for monitoring the quality of produce. Expansion of this list paralleled town growth, as the numbers of occupants of existing offices and the number of offices both rose. In 1776, the Connecticut town of Farmington (population just over six thousand) elected 206 officers: "the clerk, . . . the treasurer, seven selectmen, twenty-two grandjurymen, twenty-two listers, ten constables, twenty-five tithing-men, fifty surveyors of highways, seven leather sealers, seven weight sealers, nine measure sealers, sixteen fenceviewers, eight key keepers, seven branders, nine collectors for rates, one packer and four ratemakers."[24] This may be seen as not all a bad thing; to the extent that the towns' populations grew increasingly stratified, local politics tended to come under the sway of dominant families, and some counterforce was not unwelcome. Enlarging officialdom opened doors for some people who otherwise might not have been able to participate: "The expansion in the number of officers elected in each town, the growth of neighborhood government in the form of societies, and the increased openness of freemanship [due in part to the lowering of property requirements] all meant more opportunities for involvement in public affairs."[25]

The proliferation of town officers was a harbinger of the professionalization of government in New England. Continued growth in the region created pressures on the original town meeting structure, and more powers were delegated to officials. Since the late eighteenth century there has been a shift toward representative forms of governance. When Boston's population reached 43,000, in 1822, the city supplanted the open town meeting with a charter form of government headed by a mayor and a city council.[26] Other towns followed suit, or chose other options that weakened the town meeting tradition. Despite this tendency, however, the traditional open town meeting continues to thrive in a great many places in

New England. Its modern status will be reviewed briefly in the following section.

Modern Direct Democratic Practices

These profiles are not eulogies. Direct democracy is not just a hoary relic of the past. Its features and essential sentiments can be observed in many places in the modern world. For example, despite the assaults of industrialization, globalization, and other aspects of modernity, the legacy of Raetia lives on through its marriage to Switzerland, which is still organized into communes, makes extensive use of the referendum, and may in some ways be regarded as the most democratic nation in Europe.

Contemporary Italy likewise reflects the political patterns of its medieval past. We noted earlier the rough north–south divide in terms of civic norms and patterns of relationships that characterizes modern Italy, a contrast that is a thousand years old or more. Robert Putnam's compelling study of the connections between ancient civic traditions and modern regional government performance reveals the profound influence of the former on the latter. In 1970, nationwide reforms were implemented that devolved power to the regions and created similar government structures in all of them. The performance of the regions over the following two decades in terms of cabinet stability, budget promptness, breadth of statistical and information services, legislative innovation, bureaucratic responsiveness, and seven other measures—testified to the critical importance of civic context:

> Civic regions were characterized by a dense network of local associations, by active engagement in community affairs, by egalitarian patterns of politics, by trust and law-abidingness. In less civic regions, political and social participation was organized vertically, not horizontally. Mutual suspicion and corruption were regarded as normal. Involvement in civic associations was scanty. Lawlessness was expected. People in these communities felt powerless and exploited. . . . These contrasting social contexts plainly affected how the new institutions worked. . . . Objective measures of effectiveness and subjective measures of citizen satisfaction concur in

ranking some regional governments consistently more successful than others. Virtually without exception, the more civic the context, the better the government. In the late twentieth century, as in the early twelfth century, collective institutions work better in the civic community.[27]

The lesson for direct democracy is clear: engagement and mutualism make for better government, whatever the nature of the problems it faces. But what are the prospects for cultivating these attributes of strong democracy, as well as the others? We have no way of guessing whether they can be induced wholesale in a culture that seems as wedded to political indifference as ours, but case studies show that they can at least be induced piecemeal. There are no insurmountable theoretical obstacles to realizing the essential features of strong democracy. People can be educated into citizenship—if not necessarily in the all-consuming manner of the Greek process of *paideia,* then at least to a reasonable extent. They can confront complex technical (including environmental) problems and respond sensibly to the challenges posed, and they can sustain systems of local self-rule over long periods. Of the four examples that follow, the first two relate to the technical expertise question and the second pair to the larger issue of the modern viability of strong democratic political practices.

DENMARK'S TECHNOLOGY CONSENSUS CONFERENCES. One of the objections raised to direct democracy is that citizens allegedly have neither the inclination nor the capability to be involved in policymaking, in part because the issues are too complex and technical. This could be a showstopper with respect to sustainability, since the related issues are characterized by complexity and uncertainty. However, experience with jury trials proves, despite a mixed record, that it is not necessarily true that groups of ordinary citizens cannot deliberate sensibly over complex matters. Further evidence that bears directly on the ability of citizens to understand technical subjects (including environmental ones), to recommend thoughtful policy guidelines, to challenge experts, to pose difficult questions, and to stimulate national debate comes from the experience of the Danish technology consensus conferences.[28]

In an exercise John Dewey would have approved of, since 1987 Denmark's parliamentary Board of Technology has been organizing conferences (fifteen so far)[29] on a variety of social issues involving science and technology, including food irradiation, sustainable agriculture, and the automobile. A pool of volunteers is recruited from the general populace through advertisements in local newspapers. Candidates then submit short letters describing themselves and their reasons for wanting to take part in the conference, and the Board selects fifteen conferees designed to reflect a cross-section of Danish citizens.

The conferees proceed to educate themselves about the issue, which is generally chosen for its relevance to upcoming Parliamentary deliberations. During an intensive weekend preparatory meeting, they study a background paper commissioned by the Board of Technology. Their study raises questions, which are used by the Board to compose a panel of relevant experts and stakeholders, including (for example) unions and environmental groups. A second weekend meeting follows, during which conferees review additional background materials and perhaps ask that the expert panel be altered. The experts then prepare answers to the questions.

The final phase of the process is a four-day public forum, attended by the media, elected officials, and interested citizens, at which the lay and expert panels meet. The experts explain their views on the issue and the citizen conferees question them. The questioning continues during the second day, then the experts leave and the lay group drafts its report. The experts are allowed to review the report, though only for factual content, before the conferees present the report publicly.

These reports, which have been described as "nuanced" and "clearly reasoned,"[30] are gratifyingly sophisticated and strongly support the argument that nonexperts are able to confront complex technical issues, consider their social, ethical, and technological implications, and address them comprehensively and thoughtfully. One of the strengths of the consensus conferences is that laypeople bring a dimension to the consideration of these issues that is often missing from purely expert analyses, which reflect the viewpoints of subject-matter experts or other specialists rather than that of the broader society:

For instance, while the executive summary of [a recent study of human genome research by the U.S. Office of Technology Assessment] states that "the core issue" is how to divide up resources so that genome research is balanced against other kinds of biomedical and biological research, the Danish consensus conference report, prepared by people whose lives are not intimately bound up in the funding dramas of university and national laboratories, opens with a succinct statement of social concerns, ethical judgments, and political recommendations. And these perspectives are integrated into virtually every succeeding page, whereas the OTA study discusses ethics only in a single discrete chapter on the subject.[31]

Experts tend to represent the parochial concerns of their specialties, so it may be that proper consideration of the critical ethical dimension of sustainability will require broad lay participation of just this sort. The Danes, at least, are taking it seriously. The lay panels' conclusions are widely disseminated by the Board of Technology via subsidized local debates, videotapes, and print materials. As a result, Danes in general appear to be more knowledgeable about issues that have been examined in consensus conferences than other Europeans. While the panels do not dictate government policy on those issues, both government and the private sector take heed of the panels' sentiments in shaping legislation and business strategies.

OREGON'S WATERSHED COUNCILS. In Oregon, as in much of the Pacific Northwest, salmon are a symbol and cause célèbre, and a textbook example of the ecological links between widely separated places and habitats. Salmon are also the source of bitter economic and ideological debates between the city dwellers in Portland, Salem, Corvallis, and Eugene and the farmers, ranchers, loggers, bargemen, and others who make their living from the lands and natural resources in the rural areas.

Salmon in Oregon have been declining for a century and a half, as economic development has damaged or destroyed spawning streams and overfishing in the oceans has depleted stocks. The single most harmful project, and a showcase illustration of the trade-offs between human well-being and that of other species, is the federally owned Bonneville power system: a string of twenty-seven major hydroelectric dams in the Columbia River basin that generate enormous quantities of electricity at very low rates. The power system was celebrated by Woody Guthrie ("Roll On, Columbia, . . . your power turns our darkness to dawn") and credited with helping to win World War II, as its electrical output fed the smelters that produced aluminum for the vast fleets of American bombers and fighter planes. But the power system also transformed the Columbia River from a salmon paradise—in 1805, the year Lewis and Clark first gazed upon the mighty river, there were almost three times as many salmon in the Columbia basin (16 million) as there were people in the United States[32]—into a "bathtub river," in journalist Blaine Harden's words. Salmon thrive in the cold, clear water characteristic of the Columbia and its tributaries in their pristine state. The dams impound huge volumes of water in great reservoirs, reducing it to slackwater and allowing it to warm. They also impede salmon passage back up the rivers during the annual spawning runs, and the hydropower turbines kill countless young salmon on their way downriver to the sea. The dam and reservoirs are responsible for an estimated 80 percent of the salmon mortality in the Basin.[33]

The problems associated with the dams absorb considerable political energy in Oregon, especially in Portland and the capital, Salem. Because the Bonneville Power Administration is a federal agency, Washington gets involved, too. But the dams are not the only culprits in the salmon's decline; streams without dams are sometimes just as inhospitable to the fish. Lying in the rain shadow of the Cascades, the arid farms of eastern Oregon depend heavily on irrigation, and the intake pipes sometimes suck up fish with the river water and deposit them on the crops. Careless upland logging increases erosion and washes silt into spawning streams, burying the gravel beds where salmon spawn. Logging along stream banks reduces shade cover and raises water temperatures. Cattle herds grazing along streams can trample vegetation, to the same effect. The combined assaults of industry, farming, and ranching, and the counterpunches of river restorationists, environmentalists, and Native American groups, to whom the salmon are sacred, have created a complex and tension-fraught controversy that defies easy solutions.

A challenge like this requires many responses. One is the watershed councils, a type of watershed-based policy organization rooted in the long tradition of bioregionalist thinking about resource conservation issues. Oregon's watershed council program was enabled by 1993 legislation elaborating on an earlier bill that created the Governor's Watershed Enhancement Board. GWEB provides support and limited funding to the councils, but they are independent, local, voluntary groups. Their aim is to monitor and assess the health of local watersheds and to promote cooperative restoration efforts. Nationally, there are hundreds of such organizations, which have developed rapidly in recent years.[34] Oregon alone has more than ninety watershed councils. The councils are not, strictly speaking, governing bodies. What makes them interesting for our purposes is that they represent an interface between democratic, or quasi-democratic, practices and sustainability issues. They create forums where scientific and cultural values interact in the resolution of watershed management issues, and they depend on grassroots, bottom-up, local initiatives to work. Significantly, they tap strong sentiments about the integrity of the local environment that cross traditional political party lines to engage conservatives and liberals, Republicans and Democrats.[35]

Local support is critical. More than half the salmon habitat lies in privately owned lands scattered among hundreds of tributary stream basins around the state.[36] As elsewhere in the American West, Oregon politics is marked by conflicts between property-rights advocates and activists with restorationist and environmentalist agendas that could restrict property uses, or impose rules and regulations based on concepts like biodiversity preservation that are remote and irrelevant to many landowners, except insofar as they could interfere with making a living. Much of the land is owned and managed by the federal government, so there are several sources of possible regulation, and local landowners are sensitive about having it shoved down their throats. Rounding up local support is not always easy, and the degree of enthusiasm varies from place to place. Angus Duncan, a former head of the Northwest Power Planning Council and a veteran observer of the Pacific Northwest resource wars, puts it this way:

It's harder to move people on watershed issues the more
money there is at stake. . . . It is far harder to get those mil-
lionaire farmers [in Umatilla County, one of the state's big
agribusiness areas] to move an inch than it is to get the "red-
neck" ranchers in Wallowa County [in the rugged north-
eastern corner of Oregon] to move a mile. That is because it
is largely an economic decision in Umatilla County. . . . You
can reach people [in Wallowa County] in the same way the
wise-use movement has: through their investment in the
place and their sense of place, their cultural, historic and
family ties to the place. . . . I don't know of a watershed that
has made substantial progress if it hasn't had both the
impulse and leadership from within and the threat from
without.[37]

The watershed councils are composed above all of local people—private landowners, activists, and ordinary citizens. They may also include
representatives of local government, state and federal agencies, tribal and
other nongovernmental organizations, local industry, and academic
groups. Members are selected by local governments and are intended to
balance representation among stakeholder groups. They are thus not
democratic in the sense of being fully open to all citizens, but they do per-
form several functions intrinsic to strong democratic practice. For example, they help educate people into citizenship by involving them in watershed health issues through outreach and education, developing action
plans, conducting workshops, sponsoring school projects, and distribut-
ing newsletters and status reports. They broaden the stakeholder base and
involve stakeholders directly in the resolution of disputes that affect their
homes and communities, thereby building trust and reinforcing local
commitment. And they encourage the political willing of effects that their
political deliberations have determined to be desirable. With partial funding support from GWEB, the councils have sponsored and carried out
hundreds of watershed enhancement projects in the past several years,
including planting trees on stream banks to retard erosion, building
fences along streams to exclude cattle, helping to develop upland water
resources so cattle can be watered away from streams, building rock weirs
in spawning streams to create the habitat favored by salmon, using geo-

graphic information system technology to develop databases on water-shed health, and many others.

The watershed council approach to local resource management and dispute resolution is expanding rapidly. Oregon's program to cultivate watershed councils began in early 1994, and within about a year, more than forty councils had sprung up throughout the state. As of early 1999, as mentioned above, there were more than ninety. Their point and value, according to the authors of a study on watershed collaborative decision making, is "the assumption that these organization are means of conflict resolution, will secure ecological benefits, and maintain local control of local resources."[38] Without necessarily intending to, these writers make a reasonable strong-democratic case for this approach to resource management and ecosystem preservation:

> The lack of a sense of community may be the single most important barrier to successful long-term watershed planning. . . . A bottom-up approach to watershed planning that evolves out of particular local communities and regions is more likely to be successful in the long term or across generations than a top-down approach. . . . The initiation of community-based watershed planning combines the tools of science and technology with the social skill and hard work of people. . . . Learning how to deal with conflict is important. Better yet, learning what a community shares and has in common is the ultimate lesson of watershed planning and organizing.[39]

MODERN NEW ENGLAND TOWN MEETINGS. Despite the thinning ranks of traditional open town meetings mentioned earlier, a recent analysis found that they are still the preferred form of government in most New England towns.[40] In Massachusetts, 270 towns (85 percent) hold open meetings. In Connecticut, the figure is 90 percent; in Maine, 97 percent; in Vermont, 85 percent; in New Hampshire, 97 percent; and in Rhode Island, 68 percent. The remaining towns have shifted to representative town meetings, legislative town councils, stronger boards of selectmen, or systems splitting powers and responsibilities between meetings and councils.

Meetings are generally annual. In most towns, the primary power of agenda-setting lies with the selectmen, who issue the warrant before the meeting that defines the issues to be discussed. Voters can petition for inclusion of an issue on the warrant, however, and any matter of concern to the town may be taken up at the meeting. A moderator, elected either for a single meeting or for a short term of up to three years, controls the proceedings. Deliberations may be brief or lengthy, as appropriate. Voting may be by voice, show of hands, roll call, written ballot, or, if requested, secret ballot. Decision making is aided by various citizen committees, standing or ad hoc, which study various issues of interest to the town and offer recommendations to the meeting.

Two criticisms often made by town meeting skeptics are low attendance and poor quality of decision making. As to the first, attendance varies wildly, depending on locale, town size, the urgency of the matters to be discussed, and even the weather. Average attendance at town meetings in the six New England states in 1996 varied from less than 1 percent to 45 percent; some meetings saw 90 percent of eligible voters present.[41] In general, across all town meetings, attendance increased as town size decreased. While average attendance could be considered low, whether this is more than a theoretical concern is open to debate. Attendance at early colonial town meetings apparently was high, but eligibility was restricted to the relatively small class of freemen and attendance was compulsory.[42] By the mid-eighteenth century, with absence no longer punishable, town records show that low average turnout was becoming more common.[43] If it remains low today, there are many reasonable explanations: town growth means more issues to consider and longer meetings; greater mobility increases the number of new arrivals or temporary residents with little commitment to the town; long commutes for residents who work elsewhere may leave little time for meetings; the many nonpolitical enticements available today draw voters elsewhere; the common lack of childcare at meeting sites makes it difficult for parents of young children to attend; and so on. Despite these obstacles, attendance tends to rise sharply when a matter at issue is controversial.

As to decision-making quality, whether it is because or in spite of modest attendance, those voters who do attend are reportedly well informed. According to surveys, the quality of debate and decision is high.[44] The use of citizen committees to study particular matters in depth, especially finances, helps sustain this quality.

Traditional town meeting government has repeatedly been pronounced dead or moribund for a century or more, yet statistics suggest that it continues as a robust presence in hundreds of towns. Innovations have helped it cope with social changes and the conditions of modern life while remaining an effective mode of direct self-governance. It clearly is best suited to small jurisdictions—but perhaps that is an argument in favor of small jurisdictions as much as it is a critique of town meetings.

DIRECT DEMOCRACY IN URBAN NEIGHBORHOODS. If small is beautiful, skeptics argue, what place is there for direct democracy in a world where most people live in cities? A recent Brookings Institution study[45] examined five American cities' attempts to implement a measure of direct urban democracy. The study suggests that face-to-face democracy can be made to work reasonably well in cities, although it cannot easily be made to live up to the theoretical ideal described in chapter 5—nor, perhaps, need it do so.

The core cities study, as we will call it, dissected the governing structures of five cities with populations between 100,000 and 1 million: Birmingham, Alabama; Dayton, Ohio; Portland, Oregon; St. Paul, Minnesota; and San Antonio, Texas. These cities would seem to have little in common except for their long-term success with face-to-face democracy. They share no particular type of governing structure and vary widely in demographic composition, citizen income and education levels, and economic prosperity. However, they all launched reforms in the 1970s that decentralized governance and distributed power to neighborhoods, rearrangements that have survived the tests of time and politics. The study assessed the extent of the neighborhood associations' power and influence at city hall, the extent to which the associations were believed to represent the interests of neighborhood residents, how persistent and effective were the cities' efforts at reaching out and involving citizens in government, the relationship between strong participation and citizens' confidence in city government, and what effect all this had on policy outcomes.

The study also compared the five core cities with others that attempted similar decentralizing reforms that have not lasted. It attributed the dif-

ference in outcomes to three factors. First, motivation in the five successful cities was strong in terms of citizen demands and a clear vision from
the city leadership. Second, each system was based on small, natural
neighborhoods yet was designed to encompass the whole city,[46] and
although each city has made changes over the years, the initial system of
reforms was developed and put in place in one stroke rather than accreted
layer by layer. Third, the neighborhood associations were set up as non-
partisan groups:

> From the beginning, citizen and governmental advocates
> combined a firm refusal to let partisan politics become part
> of the participation system with an equally firm insistence
> that policy issues—and all the "politics" that go along with
> them—become part and parcel of the system's daily opera-
> tion. . . . The system is designed so that it will not be cap-
> tured by partisan politics. Because they must represent peo-
> ple from all political parties, supporters of candidates of all
> stripes, and backers of incumbents as well as their oppo-
> nents, the neighborhood organizations do not and cannot
> take stands on political candidates. . . . In general, this dis-
> tinction between issue advocacy and electoral neutrality is
> well understood and accepted by neighborhood residents.[47]

Finally, fate lent a hand: in each of the five core cities, the reformed
system got up some momentum in terms of participation levels before
running into major setbacks that eroded support for the reforms. When
setbacks did occur, the reforms had already generated enough community
confidence and support that the communities turned to the new system
for help in coping with the changes, rather than allowing the changes to
destroy the reforms.

The essential purpose of the reforms was to create strong, persistent,
comprehensive means of increasing the breadth and depth of citizen participation in local government. The intent was to give citizens the chance
to participate in policymaking at every stage (breadth), and to ensure that
those who chose to participate exerted a real influence on policy outcomes
(depth). Breadth of participation was encouraged in the core cities by cre-

ating easy means of access to the system, by giving citizens the information and encouragement needed to foster participation, and by a commitment by city governments to devote significant staff and resources to the process. Depth of participation was sought by creating means to ensure that neighborhood needs were addressed in budgeting, that the decision-making process was well defined, that citizens and neighborhood representatives had good access to city officials, and that the neighborhoods could direct the staff assigned them by the city government, so that in general they were working for the neighborhoods, not the city.

As to the effectiveness of the reforms, the bad news first: "One of the hopes in establishing structures of strong democracy is that they will increase the number of political participants. The participation systems in the five [core] cities do not accomplish this. The data are unequivocal: overall participation in the five cities is similar to that in the comparison groups."[48] Only about 16.6 percent of the population of the core cities take part in the work of the neighborhood associations. The reasons, the authors believe, have to do with competing interests and the intimidating image of open political meetings, among other factors. They also point out that levels of confidence in local government in the five core cities are relatively high—in part because of the effectiveness of the neighborhood organizations. But if success breeds lassitude, it is also true that government failure and callousness breed anger and activism. It may be a self-correcting phenomenon, and it is the structures of open democratic participation that allow the feedback mechanism to work.

The study also revealed that the poor participate much less than the middle and upper classes, and speculates about the reasons:

> Despite their openness and autonomy, it may be unrealistic to assume that neighborhood associations will overcome the class bias in American society. Other forms of political organization have also failed to mobilize the poor. The daily burdens of low-income people are powerful forces that may work to make them feel inadequate, apathetic, or alienated; such attitudes are not easily ameliorated by easier opportunities to become involved in politics. However, low-income people should not be stereotyped as having negative atti-

tudes that push them away from politics. Life circumstances, such as problems with child care and transportation, can make participation difficult.[49]

Some may find this reassuring. One strain of political theory argues that too much participation by the poor is undesirable because the poor are generally less tolerant and more authoritarian than the upper classes, and if given power would only use it to make the system less democratic. The study torpedoed that myth: the participation structures place a premium on tolerance in those who participate, and more important, most people who participate over time become more tolerant. The process both selects and educates people, *paideia*-like, into the characteristics that make the process work. If effective means could be found to increase participation by the poor, there is little reason to believe it would undermine the democratic nature of the system.

Other theoretical dangers of direct democratic practices often raised by skeptics are not supported by the core cities study. Citizens believed that the system reduced hostility, enhanced feelings of personal political power, and reduced the chances that losers in policy conflicts would try to obstruct or delay the process. City officials conceded that the participatory structures were more cumbersome than top-down systems, but "overwhelmingly felt that the benefits outweighed the costs."[50] Participatory structures like the neighborhood associations make city government more responsive, help to decrease conflict, and create a much stronger sense of community among those who participate.

NOTES

1. ". . . except for all those other forms that have been tried from time to time."

2. Anonymous, "The revision thing," *The Economist,* January 9, 1999, p. 76.

3. W. Durant, *The Life of Greece* (New York: Simon and Schuster, 1966), 254ff.

4. Ibid., 256.

5. M. Bookchin, *From Urbanization to Cities: Towards a New Politics of Citizenship* (London: Cassell, 1995), 63.

6. Ibid., 75.

7. Durant, *The Life of Greece,* 294.

8. R. Putnam, *Making Democracy Work: Civic Traditions in Modern Italy* (Princeton: Princeton University Press, 1993), 121ff.

9. Bookchin, *From Urbanization to Cities,* 97.

10. Ibid., 126.

11. Ibid., 127.

12. Ibid., 105.

13. Ibid., 106.

14. B. Barber, *The Death of Communal Liberty: A History of Freedom in a Swiss Mountain Canton* (Princeton: Princeton University Press, 1974). The subsequent discussion of the Republic of Raetia is based on this source.

15. Ibid., 100.

16. Ibid., 115.

17. Ibid., 176.

18. Ibid., 187.

19. Ibid., 201, 202.

20. J. Zimmerman, "The New England town meeting: Lawmaking by assembled voters," in *Municipal Year Book, 1998* (Washington, DC: International City Management Association, 1998), 23–29.

21. Cited in Zimmerman, "The New England town meeting," 24.

22. B. Daniels, *The Fragmentation of New England: Comparative Perspectives on Economic, Political, and Social Divisions in the Eighteenth Century* (New York: Greenwood Press, 1988), 57ff.

23. Ibid., 67.

24. Ibid., 72.

25. Ibid., 75.

26. Zimmerman, "The New England town meeting," 24.

27. Putnam, *Making Democracy Work,* 182.

28. R. Sclove, "Town meetings on technology," *Technology Review* 99(5) (July 1996): 24–31.

29. Personal communication, Mette Bom, Danish Board of Technology, April 16, 1999.

30. Sclove, "Town meetings on technology," 27.

31. Ibid., 28.

32. B. Harden, *A River Lost: The Life and Death of the Columbia* (New York: W. W. Norton, 1996), 63.

33. Ibid., 196.

34. M. McGinnis, "Making the watershed connection," in M. McGinnis, ed., "Special symposium on watershed policy," *Policy Studies Journal* 27(3) (September 1999).

35. See T. Egan, "Look who's hugging trees now" (*The New York Times Magazine*, July 7, 1996, 28–31), for a discussion of these iconoclastic politics.

36. T. Novak, "Restoration projects popping up around Oregon," in *A Snapshot of Salmon in Oregon* (Corvallis: Oregon State University Extension Service, 1998), 21.

37. Conversation with Angus Duncan, former head of the Northwest Power Planning Commission, September 5, 1996.

38. M. McGinnis, J. Woolley, and J. Gamman, "Bioregional conflict resolution: Rebuilding community in watershed planning and organizing," *Environmental Management*, in press.

39. Ibid.

40. Zimmerman, "The New England town meeting," 24.

41. Ibid., 26.

42. Ibid., 28.

43. Daniels, *The Fragmentation of New England*, 67.

44. Zimmerman, "The New England town meeting," 26.

45. J. Berry, K. Portney, and K. Thomson, *The Rebirth of Urban Democracy* (Washington, DC: Brookings Institution, 1993).

46. Except for the San Antonio participation group COPS (Communities Organized for Public Service), which encompasses about 30 percent of the city.

47. Berry et al., *The Rebirth of Urban Democracy*, 50, 51.

48. Ibid., 284, 285.

49. Perhaps it is somehow relative, rather than absolute, poverty that alienates. Poverty does not always make for political indifference. Writer Bill McKibben, for example, tells of a sojourn in the Indian state of Kerala, where average per-capita income is about 2 percent of the U.S. average, yet there is a vigorous, sophisticated, grassroots political life. The per-capita newspaper

consumption is the highest in India. Strikes and job actions are daily events. The people launch numerous local initiatives and petitions and stage frequent demonstrations. The cult of personality that dominates politics in the rest of India is a joke in Kerala. "Politics are much in the air and it is difficult to escape from them," McKibben quotes the writer K. E. Verghese. See *Hope, Human and Wild* (Boston: Little, Brown, 1995, 117–169).

50. Berry et al., *The Rebirth of Urban Democracy,* 213.

Chapter 7

Sustainability and Strong Democracy

The natural world is beautiful, bountiful, and malleable, which makes it the perfect home for an inquisitive and ambitious species like *Homo sapiens*. But it is not endlessly resilient or tolerant of abuse. It operates by rules, and within limits, which we only dimly understand and which are not always intuitive and linear in operation. Transgressing those rules and limits exposes humanity to dangers that may appear abruptly and with unexpected force. Caution suggests that we learn more about how the biosphere works and where its limits lie before proceeding recklessly in the direction of further economic growth, burgeoning populations, and badly informed and poorly monitored tinkering.

How to do that? We can make pretty good guesses, even without much additional information, about what needs to be done to ease the pace and impact of growth before catastrophe becomes probable rather than just possible. We know how to go about learning more. But most of us seem indifferent, for one reason or another, to the risks of ignorance.

At the beginning of this book, we asked the question, What is worth saving? The mania for growth that grips the modern world, even among those who already possess what earlier eras would regard as staggering wealth, is causing many precious things to be lost: species, green and wild spaces, clean air, pure water. These and other costs of converting the natural world to the manufactured world are plain to see. Less obvious, perhaps, is that the course we are on is robbing us of the opportunity to save options. These may be the most precious of all, especially to future generations, who have no voice in the present unless we give them one. A striking feature of consumption patterns among the wealthy—by which we mean not only the Bill Gateses of the world but the hundreds of millions of people in the middle class—is how careless and trivial much of it is. (Do we really need automobile air-fresheners, for instance, to lead civilized lives?) Excessive trivial consumption in the present could limit future options to vital consumption only.

Few people wish that sort of future for themselves or their children. That they fail to credit its likelihood is a problem of perception and, ultimately, of politics; that is, sustainability—which is quintessentially a dilemma of collective action—is first and foremost a political challenge. Our look at utopias shows that any plan for pursuing a vision of an ideal society is unconvincing if it ignores politics. Utopias prove that politics is inescapable.

The key question then becomes, What sort of politics would serve sustainability best? What sort of politics will most likely enable humanity to choose its future, rather than be whiplashed by ecological systems in severe or violent reaction to perturbations they can no longer absorb?

We are caught up in a system—Robert Heilbroner's "business civilization"—whose politics do not serve sustainability well. The system poses as the inevitable outcome of human nature by attributing to people certain fixed characteristics that lead logically to the behaviors of the system itself. In this circular way, liberal democratic capitalism models people as entirely self-interested utility maximizers, and not much else. The ailments of capitalist society, and our tendency to play ecological "chicken," seem inevitable if we accept the model of *Homo economicus*. The virtues and the transformative power of market capitalism over the past two centuries have helped justify this model and give it credence. But its shortcomings, failures, and injustices open it to challenge. Even economists have begun to do so, and as they probe its underpinnings they have come to think that maybe the anthropologists have a point: human beings are not well captured in *Homo economicus*. We are social creatures, not atoms, and our behavior and customs, heavily conditioned by culture as well as by the gene pool, are flexible. Any social system claiming that it alone is legitimate because its roots stem from human nature is self-delusional.

That is important because our political system (representational democracy) is deeply informed by market capitalism, which sees economic actors as linked only by the self-interested bargain, and our democracy sees political actors as linked in the same way. But this way of thinking about how people and their interactions constitute the political systems they inhabit mistakes history for fate. On the contrary, if culture is as strong a determinant of political systems as the evidence suggests, then alternatives are possible. This book has described one such alternative, strong democracy, which is based on a fuller and richer conception of human nature than *Homo economicus* and which advocates a commensurately fuller and richer politics. It is one that emphasizes the possibilities inherent in community, argues that there is a common good beyond the sum of individual goods, and asserts that both community and sustainability would be better served by wider popular engagement in the problems of governance.

We (the authors) believe in the possibilities of strong democracy, so it is with regret that we must state that its prospects are uncertain. We are unaware of any community in which the sort of idealized strong democracy described in chapter 5 has taken root. Even the neighborhood associations described in the core cities study, though they are probably the most serious contemporary attempt to implement direct democracy on a large scale, do not measure up to the ideal. They are unlikely, for example, to become the neighborhood assemblies proposed by Professor Barber without significant changes. The associations are overwhelmed with their current duties and compete with one another for resources from their city governments. They have little time or incentive to join together to consider regional or national issues.

On the other hand, a rigid idealism—utopianism—is both unnecessary and unattainable. The profiles of direct democracies sketched in chapter 6 demonstrate that many of the critical pieces of strong democracy have been successfully tested in the real world. The examples illustrate the viability of several of the key principles of strong democracy and its application to sustainability issues:

- When their decisions and input count, direct democracy engages people even though it makes heavier demands on their time and energies than merely voting. Although in ordinary times it does not radically increase the rate of public participation in government, it props open the door to participation while creating governing structures that accurately represent the interests of more than those who participate, and can create systems in which power is distributed more evenly and is less susceptible to manipulation by special interests.

- Autonomous, self-governing local communities can be aggregated effectively into larger regional or even national bodies.
- A robust, tenacious, and deep-rooted tradition of civic engagement and self-governance can be created and sustained for long periods. People can be educated into citizenship, especially by participation.
- Ordinary people without special technical expertise can understand complex environmental issues and can contribute thoughtfully, and in ways that represent broader interests than those of specialists, to social decisions about those issues.
- People with varying interests and viewpoints can come together in a political setting and begin to will a common environmental future.

These are hopeful lessons. As to whether strong democracy would make for more sustainable economies, there is nothing here to suggest that our earlier assertions are wrong, and some evidence, such as the accomplishments and promise of the technology consensus conferences and the watershed councils, to suggest they may be right.

A definitive answer to the question is beyond the scope of this book and would require, as scientists interminably insist, further research. Portland, one of the focuses of the Brookings Institution core cities study, is one of the most environmentally progressive cities in North America and provides at least anecdotal support for the idea that strong democratic structures are conducive to progressive environmental policies. To achieve the same effect in other large cities and towns, where people tend to be more isolated from the realities of resource-based economic policies than rural folk, may require that their education into citizenship concentrate acutely on those issues. It would be unwise to make too much of Portland's happy combination of political engagement and environmental progressivity. Portlanders may be more sensitive than most urbanites to environmental and sustainability issues because the city's spectacular natural amenities have attracted many environmentally conscious people there in the first place, and because of the way resource conflicts are continually brought home to them in state politics and the tensions between urban and rural interests.

In the United States generally, the times may not be especially conducive to the spread of direct democratic practices. Robert Putnam, for one, has documented the decline of social and civic capital in the United States in recent years in his well-known article "Bowling Alone."[1] People are voting less, attending public meetings less, working for political parties less. They trust the national government less. Churchgoing has declined, as have union membership and participation in parent-teacher associations. Volunteerism is waning. Fraternal organizations and women's groups report serious drops in membership. In reference to the title of Putnam's article, while more Americans than ever are bowling, the incidence of bowling in organized leagues dropped 40 percent between 1980 and 1993. Even informal socializing with neighbors has declined. Although membership in what Putnam calls "tertiary organizations," such as environmental groups, is up, affiliation with such groups rarely involves the same kind of face-to-face connectedness and common work or socialization as traditional civic groups. The reasons for the decline are not clear, though certain demographic changes and television—which has displaced social forms of leisure with private ones—are leading suspects.

Students of civic health disagree about whether Putnam is right.[2] If he is, it is a sad development, and it does not seem that much good can come of it if one believes, as we do, that involvement in community is one of the fundamental wellsprings of human happiness and fulfillment, as well as a key to sustainability. A ray of hope is that there seems to be a widespread longing for community, so well recognized that it has become something of a cliché in analyses of American society. People sense that something is missing, and this may create an openness to experiments designed to restore the loss. As the core cities study demonstrated, political engagement is superb at promoting a sense of community.

But people still have to decide to do something about it. They have to begin the willing of a vision through personal action. Intelligent systems thinking about how to create sustainable communities can make a big difference but can go only so far. Ethan Seltzer, a professor at

Portland State University who has studied these issues for years, puts it this way in the context of solving transportation problems for sustainability:

> One of the things we've learned in this region . . . is that even if you arrange things in space as best you can—building orientation, pedestrian amenities, all the tricks you can think of to encourage people to get out of their cars—it's not enough unless people actually change their usage of the transportation system. Fundamentally, you can't get there unless people change their behavior. If people really expect to "walk their talk," that means they're going to have to change their behavior. Otherwise what they're doing is walking the talk [partway] and driving the car the rest of the way.[3]

Beyond the simple yet important insight that personal choice matters, we have no particular solutions for this dilemma of declining social capital. It is a challenge that faces us all. It may be that no more than 16.6 percent of the populace will ever be seriously engaged in participatory politics. The modern world is full of meaninglessness for many people, but it compensates with distractions. Dutch historian Johan Huizinga saw humanity as *Homo ludens* (man the player), for whom the search for distractions is compelling. We play games. A stockbroker like Michael Milken who makes half a billion dollars in one year trading junk bonds on Wall Street is really not just making a living. For others, it's Nintendo, bridge, baseball, or following the soap operas. For 16.6 percent, perhaps, it's politics. The citizens of the Athenian *polis* or the Raetian communes arguably had fewer distractions, which may have contributed to their evidently more thorough politicization. Chances are we'll never have that—unless the dire predictions of environmental catastrophe come true. In that case, don't bet against a dramatic rise in political consciousness. In the spring of 1999, OPEC nations decided to curb exports to boost oil prices. As the cost of a gallon of gasoline nearly doubled on the U.S. west coast, the Internet was flooded with outraged e-mail messages. Somebody proposed a protest, a one-day "Gas Out" when motorists would boycott all gas stations.[4] Given how pampered American motorists are in general, let alone Los Angelenos, no one should imagine that any less-trivial environmental assault on our lifestyles will be accepted gracefully. There is a great deal of latent political energy out there waiting to be channeled.

All political systems generally have this kind of slack in them. Never, under any political system, is everybody always engaged in politics.[5] This is a reason why relatively low numbers of participants may not be all bad; it means that there are untapped resources available when the system is perceived to be no longer effective and a change is necessary. This is especially true when the options for responding to a seriously failing political system have become restricted to what economist Albert Hirschman calls voice (protest, agitation, and other constructive expressions of dissatisfaction).[6] The other option is exit (bailing out). The United States is a nation built by exiters (emigrants), many of whom found that the voice option was closed to them by harsh oppression. However, with the physical frontier closed more than a hundred years ago and the consumerism frontier shrinking rapidly, voice may be the best option for Americans in future times of stress. (And not only for Americans: with the world getting fuller day by day, increasingly there is nowhere to run, for anybody.) Voice can be given expression by mobs and rabbles, or it may find a bounded forum that allows for the full expression and resolution of conflict through the political arts. Strong democracy is just such a forum, perhaps the best one for the fullest expression of many voices with a stake in the future. When prosperity can no longer displace politics á la the Madisonian dream and we must face the inescapable tests posed by the declining sustainability of our current way of life, let us hope that the tools of strong democracy will lie close to hand, ready to be taken up and used to build a better world.

NOTES

1. R. Putnam, "Bowling alone: America's declining social capital," *Journal of Democracy* 6(1) (January 1995): 65–78.

2. See, for example, R. Morin, "Who says we're not joiners?" *The Washington Post*, May 2, 1999, B5. And perhaps the sociopolitical climate is changing in ways conducive to a sustainable, civic-minded society. The population cohort known as Generation X (those Americans born between 1965 and

1978) may be more attuned to environmental activism, shared sacrifice, and direct democratic practices than the Baby Boom generation that preceded it. See T. Halstead, "A Politics for Generation X," *The Atlantic Monthly,* August 1999, 33–42.

3. Conversation at Portland State University, September 5, 1996.

4. B. Moore, "Adding fuel to the ire," *Los Angeles Times,* April 12, 1999, D-2.

5. Even in Periclean Athens, if the numbers we cited are accurate, typical attendance at meetings of the *ekklesia* was perhaps 5 or 10 percent of the eligible electorate.

6. See A. Hirschman, *Exit, Voice, and Loyalty: Responses to Decline in Firms, Organizations, and States* (Cambridge: Harvard University Press, 1970).

Index

amazon.com. 1 Centerpoint Blvd
P.O. Box 15550
New Castle, DE 19720–5550

A. Price
n Road
NY 10536
A

Charlotte A. Price
37 Garlen Road
Katonah, NY 10536
USA

ISL

chag16806/-3-/6186/std–us/1676008/914 232-8380

Description	Format	Our Price	Total
Hawken, Paul	Paperback	$14.36	$14.36
Prugh, Thomas	Paperback	$22.50	$22.50
Daly, Herman E.	Paperback	$25.00	$25.00

Subtotal		$61.86
Shipping & Handling		$2.97
Promotional Gift Certificate		$-2.97
Shipment Total		$61.86
Paid via Visa		$61.86
Balance Due		$0.00

ve you the speediest service possible. The other items in your order
s order will not exceed the amount we originally promised.

om the "Your Account" link on our homepage.

.com, and please come again!

Earth's Biggest Selection

Lexington, KY 40511-1013

Items that are returned more than 30 days after delivery, are in unsellable condition, or are missing parts, will be charged a re-stocking fee at our discretion.

* If you are returning an item for one of the following reasons, please contact us at returns-problems@amazon.com:

☐ The item is missing parts/accessories.

☐ The product became defective/damaged after it arrived.

☐ You are returning an item that was delivered by a specialty shipper (e.g. Eagle or NationStreet).

Note: To return a gas-powered item, please contact the manufacturer directly.

For other questions or issues with your order, please visit our site first and click the Your Account button. In Your Account, you can access many account maintenance features, such as viewing the status of your orders, canceling unshipped orders, or changing your e-mail address or password. If you need additional assistance, please contact us at orders@amazon.com.

Thanks for shopping at Amazon.com!

amazon.com®

Returns are easy!

For most returns, there is no reason to contact us. *
Simply visit the Returns Center on our Web site at
http://www.amazon.com/returns. Here are the details:

- Within 30 days of receiving this shipment, you may
  return any book in its original condition (or that we
  recommended and you didn't enjoy), any unopened
  media (CD, DVD, software, video game, etc.) or any
  other item in new condition with original packaging and
  accessories for a full refund of the cost of the item(s).
  We'll even refund the shipping cost if the return is the
  result of our error.

- You can expect a refund in the same form of payment
  originally used for purchase within 7 to 14 business
  days of our receiving your return. If you are returning a
  gift, you will receive a gift certificate for the value of
  your return, which may be used any time toward your
  purchases at Amazon.com. Please note that we cannot
  exchange items (unless they are defective or damaged).

If you cannot access the Returns Center on our Web
site, please fill out the form to the right, include this
entire packing slip with your return, wrap the package
securely, and send it to the address below. For your
protection, we recommend that you use UPS or Insured
Parcel Post for shipment.

Ship to: RETURNS CENTER - AMAZON.COM

Did you receive this order as a gift?

☐ Yes   ☐ No

If so, please know that we will not send the gift giver
confirmation of the return, and you will receive a credit
for your return in the form of an Amazon.com gift
certificate to be used toward any future purchases.

Please choose the reason for your return:

☐ Product was defective/damaged when it arrived
  (please indicate below if you want a replacement).

☐ Product performance/quality is not up to my
  expectations.

☐ I ordered the wrong item.

☐ Item took too long to arrive; I don't want it any
  more.

☐ No reason, I just don't want the product any more.

☐ Product is not fully compatible with my existing
  system.

☐ Amazon.com sent me the wrong item (please
  indicate below if you want a replacement).

☐ I found better prices elsewhere.
  If so, where? _____

Comments:

amazon.com.

http://www.amazon.com
orders@amazon.com

Charlot
37 Gar
Katonah
U

Amazon.com
1 Centerpoint Blvd
P.O. Box 15550
New Castle, DE 19720–5550
USA

Your order of November 18, 2000 (Order ID 103–5075836–7464656)

Qty	Item
	In This Shipment
1	Natural Capitalism : Creating the Next Industrial Revolution (1–6–4)
1	The Local Politics of Global Sustainability (1–6–4)
1	Ecological Economics and the Ecology of Economics : Essays in Criticism (1–6–4)

We've sent this portion of your order separately at no extra charge to
are shipping separately, and your total shipping charges for

You can always check the status of your orders

Thanks for shopping at Amaz